Albert Alberg

The floral king

A life of Linnaeus

Albert Alberg

The floral king
A life of Linnaeus

ISBN/EAN: 9783337271619

Printed in Europe, USA, Canada, Australia, Japan

Cover: Foto ©berggeist007 / pixelio.de

More available books at **www.hansebooks.com**

THE FLORAL KING:

A LIFE OF LINNÆUS.

BY

ALBERT ALBERG,

AUTHOR OF

F bled Stories from the Zoo," "*Gustavus Vasa and his Stirring Times*," "*Charles XII. and his Stirring Times*," ETC., ETC.

LONDON:
W. H. ALLEN & CO. 13 WATERLOO PLACE.
1888.

LONDON:
PRINTED BY T. BRETTELL AND CO.,
RUPERT ST., HAYMARKET, W.

PREFACE.

N presenting this unpretentious little volume to the public, I must express my sense of indebtedness to the late Dr. EWALD ÄHRLING, of Arboga, in Sweden, who devoted upwards of twenty-two years to zealous researches in the vast Linnean archives, at home and abroad, and from whose published selection of Linnæus's correspondence the extracts met with in the following pages have been chosen, with a view of making the reader familiar with

Linnæus's mode of working, his home-life and surroundings.

Nor must I, amongst other Swedish sources, omit to mention the charming biographical poem, entitled "Blomsterkungen," by Dr. Herman Sätherberg, of Stockholm, which first suggested the idea of gathering a few flowers in the garden of Linnæus's child life, and which, transplanted into English soil, may be acceptable to those who recognise the aphorism that "the child is father to the man."

ALBERT ALBERG.

London, *October*, 1888.

CHAPTER I.

HERE grew a large and stately linden tree of yore in the parish of Stenbrohult in the southern province of Småland in Sweden. It stood near the boundary of two parishes, and towered like a landmark in the surrounding woodlands, the trees of which were comparatively dwarfed by its imposing presence. This tree must have been venerated by many, and loved by some, for several lads from the parish when leaving home, and starting upon their career in the schools, took themselves surnames which they derived from their

cherished linden tree. One called himself Lindelius, another Tiliander, and so on. One poor young rustic, Nils Ingemar's son, from the neighbouring homestead of Jonsboda, was sent to school at the expense of his uncle on his mother's side, the Swen Tiliander anent, and who had now become rector of an adjacent parish church. This young boy thinking of his well-beloved linden tree, under whose branches he had been wont to hold sweet converse with nature, and through nature with his God, now on starting in life with hopes of becoming a pastor himself in his own parish,—the laudable ambition of the sons of well-to-do peasants common to this day in Sweden;—this boy assumed the name of Linnæus.

Swen Tiliander had in his youth lived for some years in Germany, and there acquired a love for flowers and some knowledge of horticulture. When he became rector of a rural parish in Småland, he found great enjoyment in planning a pretty little garden at his rectory, and stocking it with rare and beautiful flowers from abroad. In this garden young Nils Linnæus spent many happy hours, and caught his uncle's enthusiasm for the culture of Flora's children; and no sooner had he struggled through his academic period, subsisting on slender means, and reached the goal of his ambition, that of becoming curate in his native parish, than he hastened to follow his excellent uncle's example, and planted a small garden at his little cottage Råshult, allotted to him for a residence. To this unpretentious little home, which boasted of no luxury save the floral surroundings, Nils Linnæus in 1706 brought his own rector's daughter, Christina Brodersonia, from the neighbouring manse as his happy bride.

Here dwelt Love in a cottage, embowered by all the darling flowers loving hands could plant and tend. The young wife was equally delighted, as she had never before seen a garden, for in the sterile and stone-bound soil of Småland, amongst its forests and waters, few gardens to this day are to be met. When autumn came and despoiled the plantation of its manifold and variegated beauty, her young heart grieved that the long Swedish winter would intervene before spring would gladden her again by unfolding to her its choice floral treasures, and she felt quite impatient with cruel Father Winter for ruthlessly nipping her beautiful flowers with his early frost.

At last spring returned, and what joy did it not bring to her yearning heart; for not only is spring in Sweden the most beautiful season of the year, when nature in a few days wakes from wintry sleep from under the snowy cover, and the soil gratefully absorbs the remaining snow to fertilize the earth, whole masses of ice, dissolving into water, hasten away in merry little rills, as if afraid of being hid in the earth, and rush to swell the tributaries of the many rivers, which all make for the cool, clear sea—and when every twig and frond is covered with eager budding leaflets, kissed to life by spring, and inquisitive to look abroad—at this delightful season, when all nature rejoices at the spring-time of her new existence, "just when the cuckoo with mystic notes heralded the advent of the floral season," the curate and his young wife, on the 13th of May, the old Gregorian style, anno 1707, were supremely blest by the seasonable advent of a young cherub, for to them was that day born a son and heir—and alighting upon earth, as he did, in the joyous, verdant spring, in such a happy, floral home, it seemed as if the pretty little flowers of the curate's garden had enticed him there from the first to become their playmate, and subsequently to become their most ardent lover.

This eminently pastoral home was situate near the rural church, which stood surrounded by pleasant fields and meadows. On its western side mirroring its tall spire in the placid waters of the Möckeln, a creek of which large lake ran for some considerable length, until it reached the site of the church. To the south lay glorious beechen-woods, and the landscape was framed by high mountains in the north, clad with a stately fir and pine forest, stretching eastward.

The young pastor could show more than 400 species of exotic flowers in his garden, which in those days were deemed a most curious and rare collection. In the centre of his plantation he had made a high and circular flower-bed, which he designated the banquetting table, and on it grew the choicest flowers, cunningly arranged to represent various dishes, and round this were grouped a number of shrubs and small detached beds, disposed to represent the guests come to join in the floral feast, a quaint and altogether living group. The father, who loved his little son dearly, was wont to deck his cradle with pretty blossoms, and when the little fellow cried—as young denizens will do—the mother invariably quieted him by simply putting a flower in his hand, for it exercised a perfect spell over his infantine mind, and served to draw his attention to the choicest plaything Heaven has given a darling child.

The young couple removed the following year to the neighbouring rectory of Stenbrohult, to which living Nils Linnæus, on the demise of his father-in-law, was appointed, and to which place he, with few exceptions, succeeded in removing his choice collection of flowers and shrubs into a newly planned garden.

The little boy had been baptized "Carl," and when so far advanced that he could toddle by his father's side into the garden, he did so every day, admiring and loving the floral pets. When he grew a little older he had a few beds allotted to him, where he cultivated at least one plant of every species that grew in the garden, and when only four years old he readily learned the Swedish names of all the common flowers, besides all his father could tell him about their qualities, often puzzling his paternal teacher by quaint and striking questions, revealing a reflecting and original mind. And here with his little implements he eagerly took to Adam's original occupation of gardening, and worked at it to his little heart's content, and to the great delight of his flower-loving parents.

His mother, in particular, early destined him, in her heart, for the Church, as so many of their relatives had chosen that career, and it was with deep concern the parents, as the lad grew up, noticed his aversion from reading but that which related to Botany and Natural History; that he was for ever searching fields and woodlands in quest of flowers, and his loving mother complained that no sooner had he got a new flower than he cruelly pulled it to pieces, for the little fellow loved to penetrate, as far as it was possible, into the secrets of nature. People called him playfully "the little Botanicus," for he knew where grew every kind of herb and plant in the neighbourhood, and he assiduously collected and dried all the specimens he could get, and frequently littering his mother's room.

One day his mother found that he had even appropriated her much treasured Bible to press some new-found flowers in, and she began gently rating him for this.

"Dear child," she said, "you must not put herbs and flowers in my beautiful book. It would be quite a sin to spoil the Holy Bible."

"Pray forgive me, mother, but these are the most beautiful flowers I have ever seen, so I thought I would preserve them the best of all, for I have heard both you and father say that the Bible is the Book of Life, and surely if I put the flowers between its leaves they would retain their colour, the Bible keeping them alive for ever."

"Child, when we call the Bible the Book of

Life, we mean not by that the life we see before us, but the spiritual growth of our souls, for every thought we think is a flower culled in the garden of our soul. There, like on earth, grow many various plants, some of wondrous beauty, and others stained with sin. But every time we humbly read in the sacred writ a seed is sown in our heart which some time will bloom and bear holy fruit."

"How beautifully you talk, mother."

"Well, you must diligently read your Bible, and in your heart will grow the seed of goodness and humility, but I fear——."

"What do you fear, mother?"

"I fear you love the fair flowers of the earth too much ever to care for the seeds that were watered with tears in the Garden of Gethsemane."

"O mother, no, I won't forget my Bible. But when I see a flower I think this way:—Why does God make the cold, damp earth grow such lovely creatures, with such beautiful colours? Why, if not to make us happy with the sight? And then I almost fancy the flowers saying with their petal lips, 'Look at us, and think how kind and good is God!' O, mother every flower *must* have been a thought by God!"

"Why, how you speak, child. Well, yes, you are right, it must be so. We are all children of the same God."

One day when his parents had scolded him for busying himself with his favourite occupation of culling and drying flowers instead of attending to his lessons—for botany was very little thought of in those days—the poor little culprit repaired with tearful eyes to a favourite place on a sunny hill, where grew a profusion of wild flowers, for naturally he felt as if he would unbosom himself to those innocent little friends for whose sake he had been chided. The tears trickled down his cheeks, and he folded his hands in prayer as if to ask God to help him to understand why he must be scolded for loving the beautiful flowers too much, all the more, as from his very babyhood he always had been encouraged in tending them. Weary of trying to unravel the knot of man's inconsistency, he lay down amongst his pets, and looked through his tears at their loveliness until their manifold colours blended shimmering before his eyes, and with a smile, reciprocal of their love, he fell asleep.

The fragrance of the hill-side flowers brought to his mind's eye a beautiful vision in his lulled sleep—a glorious dream, the embryo-conception of that thought which in after years was begotten, and with the flash of genius burst upon the learned world to revolutionize botany with the demonstration of the "sexual system" of plants.

In his sleep he was conscious of lying on a sunny hill surrounded by thousands of wild and redolent flowers, but he saw how every flower liberated itself from its stem, and that the petals took the shape of wings, and in the endless chain that moved around him in airy motion, as in a graceful dance, trembling with delight, he plainly discerned alternate male and female features gaze lovingly at him from out every chalice, and they sang to him a song of exultation, praising the love of the Creator; they spread their fragrance as incense in token of holy adoration, and their balmy breath whispered to the sleeper that in return for his ardent love for them they had invited him to this their wedding, which occurs but one time in the life of every little flower; that they had made him their king elect, and that when he had grown up and become a man, and had brought proper order and harmony into the three kingdoms of nature, man should also call him, and recognize him, as the "Floral King."

❋ CHAPTER II. ❋

AVING reached the age of ten, Carl was sent to school at Wexiö, the diocesan capital of the province of Småland (with an Episcopal See). Wexiö was then noted for the study of Latin and Greek with a view of preparing its scholars for the clerical profession, and as Carl was destined by his parents to become a clergyman he was sent thither. But in former times teachers in Sweden laboured under the perverted idea that learning was to be driven into the disciples; and few there were who understood how to evoke an

ardent desire to acquire knowledge. Our young hero endured rough usage at the hands of the teachers, for he felt no interest in any subject except botany and natural history, subjects which were lamentably neglected at this collegiate school. Carl was pronounced a dunce, but when the short Swedish summer came round on its annual visit to Mother Earth, young Linnæus received new life, and he scoured hills and vales in pursuit of his favourite study, at nature's own breast, and during the summer vacations he explored the whole long line of forty English miles between Wexiö and his rural home, and stored his herbarium with hundreds of rare plants. Thus passed eight years, when his father, during one of the terms, journeyed to Wexiö to consult with the teachers respecting his son's future career. The unanimous verdict of the learned fraternity was, that young Carl Linnæus had neither ability nor aptitude for learning, and they one and all freely advised the father to expend no more money in keeping him at school, but to put him to some trade at once, as a joiner or tailor. The poor pastor was deeply grieved, for it completely crushed his and his good wife's long cherished hope of once hearing Carl preach; and what heart-burning this decision caused in the breast of the young ardent botanist might easily be imagined; this cruel blow dealt by incapacity (incapacity of discernment common in learned men of narrow spheres and narrow views) to a young, penetrating and soaring genius.

In this great distress Pastor Linnæus called upon a friend, Dr. Rothman, a physician in Wexiö, who also taught physiology and botany in the school. His verdict however was, "Well, a preacher Carl certainly never will be, but he might become a famous physician, and that profession will feed a man as well as the Church. Your son is far advanced in natural history and, without gainsaying, the foremost scholar in botany. If you will permit, I'll take him into my house; he shall eat at my table gratis, and I will myself read with him during the year that remains before he can proceed to a University." It need not be told how gladly father and son accepted this generous and well-timed offer.

Carl now removed to Dr. Rothman, and this learned gentleman with great discernment made it clear to his protegé of what great advantage, and how indispensable were Latin and Greek for the study of medicine, botany and natural history.

The dead languages now became endowed with a new living interest, and instead of *Justinus's Martial Narration* and *Cicero's Orations*, he studied with avidity *Pliny's Natural History*—performing thus a double study at the same time; and he became so enthralled with his author that the noble Roman's simple and pithy style of expression, as a natural consequence, unconsciously became inherent in his Scandinavian disciple, though a thousand years separated them. Thus true genius ever bears fruit in remotest ages.

Dr. Rothman grew daily more and more attached to his pupil, who made amazing progress, and whose transcendent genius became more and more evident. He found great delight in guiding the young naturalist in his studies, but soon found, with little surprise and no envy, that the pupil far outstripped his teacher. Linnæus could acquire no more from him, and besides, the young student had already carefully and practically studied those herbs and plants, insects and animals, which were to be found in his neighbourhood.

It was now high time for him to proceed to the University, when new troubles arose. His father had consented to Carl privately studying medicine and botany, under Dr. Rothman; but he still cherished the hope that, once arrived at the University, he would seriously apply himself to the study of Theology. As yet Pastor Linnæus had forborne to tell his wife that their son wished to become a physician, as he knew that would grieve her even more than "if he were to change religion." However, it could no longer be kept secret from her: she was quite inconsolable, for the medical profession had very little *préstige* in those days, and she felt great remorse at the love she herself had borne the flowers, even before her son had been born, and she put all the blame on the innocent plants, and sternly forbade her second son, Samuel, ever to devote himself to such a useless and unprofitable study as that of herbs and flowers.

At last, after much importunity, and even tears, Carl gained the day; and nothing now remained but to get the certificate from the school. It was framed in very quaint and significant words, and it is curious that the trope of a tree, carried all through, should have been applied to the future Professor of Botany. It read as follows: "The youths in schools may be likened unto young saplings in a plantation, where it sometimes happens, although seldom, that young trees—despite the great care bestowed upon them—will not improve by being engrafted, but continue like wild untrained stems, and when they are finally removed and transplanted, they change their wild nature and become beautiful trees, that bear excellent fruit. In which respect, and no other, this youth is now promoted to the University, where, perhaps, he may come to a clime that will favour his further development."

With this questionable recommendation Carl Linnæus went to Lund, the southern University of Sweden, in 1727. His parents thought it desirable that he should proceed to this seat of learning, as a distant relative, Dr. Humærus, was resident Dean of the ancient Cathedral of that town. But the first sound that met our young traveller on his arrival was the funeral knell of that reverend man, who at that hour was being carried to his grave. However, fortune befriended Carl, for he met his old private tutor, Magister Höök, and he saved him from producing the humiliating school-certificate, which no doubt would have been to his detriment, and, instead, presented him for matriculation at the University as being his own private pupil.

Linnæus boarded and lodged at the house of Lector Kilian Stobæus, who lectured in the University on natural history, geology and botany and was a man of acknowledged profound learning in these sciences, and who possessed a large private collection of stones, shells, birds and dried herbs. At this house also lived a German student of medicine, David Samuel Koulas, eight years Linnæus's senior, who had the use of Stobæus's library, and who took upon himself to secretly lend his young comrade what books he required in botany. The old mother of the learned host had observed that a light burned in the small hours of the night in Linnæus's room, and being fearful lest the young man might forget to put out his candle on going to bed, and thus endanger the whole house, she told her son of this her surmise, and cautiously one night he went up stairs to Linnæus's room to surprise the negligent fellow, but he was himself surprised, for he found the young student, in the dead of night, assiduously occupied in comparing the varying opinions of the greatest botanic authorities of his time—in earnest and devoted study of his favourite science. This discovery won the affection of his teacher, and Stobæus from that moment gave him the free use of his library, also the keys of his collections; and, like Rothman, took a sincere interest in the gigantic strides the young naturalist made in his science. Stobæus even offered to make him his heir, though he himself was then not more than thirty-eight years old, and the following year became professor, and in all likelihood had the prospect of a long and useful career before him. Linnæus diligently collected all species of herbs and flowers from the neighbourhood of Lund, but during the summer vacation he visited his home, when, after consulting his friend Dr. Rothman, it was decided that Carl should change to the University of Upsala, where the far-famed Professors Roberg and Rudbeck lectured in theoretical medicine and botany, and where a much larger library and the fine botanical garden afforded greater facilities for the study of his favourite science.

When a child he had heard of the wonders of this plantation of beautiful flowers, and in his childish heart he had wished he might one day become the gardener of that place. With joyful expectations he now set out on the journey with this *Eldorado* for his goal, there to pursue his studies in which centred all the ardour of his soul—the yearning to read with inspired eyes the writ of love and perfect order and harmony which God has traced everywhere in His creation, but which only those who are supremely blessed with virtue and genius are allowed to decipher.

❀ CHAPTER III. ❦

NE of the most eager botanists of the seventeenth century was no doubt a learned German physician, Joachim Burser, resident in Annaberg in Saxony, for he had wandered through Germany, Switzerland, Italy, and the South of France, and climbed the Pyrenees, gathering the flora of each land. In 1625 he was appointed Professor of Medicine and Natural History at the University of Sorö, in Denmark, where he taught till his death, in 1649. He had given his duplicate plants to his learned friend Caspar

Bauhinus, and this erudite botanist had also access to the whole of Burser's collection, and from this splendid herbarium wrote his famous book, *Pinax theatri botanici*, published in 1623. After Burser's death the whole of his collection, a herbarium consisting of twenty-five large volumes, was brought to the Sorö University, from which this valuable and unique treasure was taken as war booty in 1658 by the Swedes, and brought over to Sweden. It finally fell into the hands of a Swedish nobleman, by name, Coyet. Professor Olof Rudbeck, *Senior*, of Upsala, hearing of this rare collection, persuaded the possessor to turn it over to Upsala University, where it would be put to great use in the study of botany. This was done, [and in 1666 the northern University congratulated itself upon the possession of this singular spoil of war. The learned Professor Rudbeck again put the collection to great use, and wrote a highly meritorious and gigantic work the *Campi Elysii*, in eleven tomes, which contain more than 6200 botanical drawings. This erudite work entitled him to the honour of being the greatest botanical authority at the Universities of Scandinavia until the advent of Linnæus.

It was in order to avail himself of the study of these learned works and famous herbarium, that Carl Linnæus, in 1728, changed to the University of Upsala. But the also famous Professor Olof Rudbeck, *Junior*, was already old, and had ceased lecturing on Botany and Natural History, and living much within himself, was almost inaccessible to strangers.

To this disappointment for Linnæus was added his great stress of poverty, for he only possessed the small sum of money of 100 daler in silver (about £2), the last help his father could bestow on him, and he had no prospect of obtaining a situation as tutor in a private family—which was a customary means by which poor students sustained themselves while studying at the University —for he had chosen medicine as a profession, and that branch had very little honour in those days, even at Upsala. He was compelled to get into debt to procure himself food, and became so reduced in circumstances that he could not even afford to get his shoes mended, but had to put paper in them in lieu of getting them soled, and his whole dress bore evident traces of extreme poverty. But still he refrained from addressing himself to his friend Stobæus, now professor in Lund, for he had left that University, for Upsala, without first taking counsel of him, and he very naturally felt that he would rather endure all kinds of privations than apply, in his emergency, to that gentleman for succour. And during a whole year poor Linnæus endured great physical wants for the love of his favourite science. A crust of dry bread with water was, during this time, generally his breakfast, but he was up betimes, at early cockcrow, arranging his notes, and making preliminary outlines for a new great reform in Natural History; and with clear and steady aim he gradually revealed, link by link, the great chain of things created, which to most men's eyes lies entangled in a bewildering maze, for Linnæus's genius particularly adapted itself to discerning the order and harmony of creation. In these sacred hours of deep study, when by the light of his genius, with devout feelings he read the Book of Nature, which his diligence had unsealed, he felt himself supremely happy beyond the measure of most mortals. His mind and eyes, strained by assiduous night work, always found an additional smiling reward when he took a walk amongst the beautiful daughters of Flora in the Botanical Garden.

In the autumn of 1729 a letter arrived from his father, earnestly and lovingly entreating him to leave Upsala and return home to try and prevail upon himself to take clerical orders. The great privations he endured did not set him against pursuing his favourite study, but he thought it due to filial obedience to follow his father's earnest desire, seeing that through his poverty he was unable to remain long enough at the University to properly develop his great faculties. With lingering steps he repaired once more to the beautiful botanical garden to bid farewell to the place he loved so well, and to gaze for the last time upon the darling children of Flora, which from many distant climes had been transplanted thither.

The early ravages of autumn had already begun, and here and there lay a withered flower or a broken stem, emblems of his own withered hopes and broken career, and he slowly passed on as one who dreamily walks past where his life's love lies crushed in hopeless death.

Amid the spoliation around he suddenly discovered a new exotic flower that for the first time was in bloom. With eager steps he approached it to admire and examine the rare floral treasure, and to gather it for his herbarium as a memory of his parting visit, and he was in the very act of cutting it off when a stern voice from behind reached him, peremptorily forbidding him to touch the flower. He turned round and saw an elderly clerical man, of venerable appearance, who asked him by what right he presumed to gather flowers in the Botanical Garden. Linnæus excused himself that he had wished to take it with him as a memory of the place he now was compelled to leave for ever. In the course of conversation the inquirer learned who Linnæus was, and the sad cause of his leaving the University; and the old man soon perceived that this student was endowed with more than common abilities. He was pleased to hear that he devoted himself to the study of botany, which was also the old man's favourite theme, and which likewise had caused him to visit the Botanical Garden when but just now he had returned home to Upsala, after a whole year's absence in the capital. This kind man, who was the venerable dean, Olof Celcius, *Senior*, requested Linnæus to accompany him home, and when Linnæus had fetched and shown him his fine herbarium, and he had learnt more about him, he gave the poor, struggling botanist a room in his own house, with free invitation for Linnæus to take all his meals at the dean's own table; and he also allowed him free access to his library, and told him to assist him in making a catalogue of the plants growing in and around Upsala. He also promised Linnæus that he should be allowed to accompany him on his journeys. In one word, Linnæus was saved to science, saved through the instrumentality of an humble flower; that flower had a mission from the floral community at large, to repay Carl for his ardent love for them, and to be the turning point whence he should start on a continuous and glorious career.

Being the *protegé* of the venerable dean, several young students soon began to take private instructions from Linnæus, who thus was soon enabled to get himself both clothes and shoes. A word from his patron sufficed to open the doors of the aged Professor of Botany, Olof Rudbeck, *Junior*. And when, towards the end of 1729, Linnæus had written a treatise upon botany, inspired by the new light which his genius threw upon science, Dean Celcius showed this to his friend, the old Professor of Botany, who from that moment became Linnæus's particular patron, and likewise allowed him the free use of his library, with particular permission to study the learned professor's own manuscripts, and his exceptionally fine drawings of Swedish native birds; added to which he was appointed tutor to Rudbeck's young sons. Dean Celcius,

Senior, was known also as a poly-historian, and at this period was busily engaged in writing a book of great erudition about the trees and plants mentioned in the Bible.

The little treatise which so captivated the old learned professor only consisted of a few pages, which Linnæus, as a mark of respect, tendered his patron, Olof Rudbeck, on the 1st of January, 1730, but which like a flash of lightning revealed the mystery that shrouded from man's eye the secret that herbs and plants generate in much the same manner as do the animal world, of which grand truth Carl Linnæus had long had a foreshadowing, and now, after close observation by the light of his genius, it at last stood clear and demonstrated before him.

Linnæus's sexual system of grouping the plants according to their means of propagation at once revolutionized the accepted classifications in botany, and is perhaps the work bespeaking the greatest penetration of his felicitous genius. He was not yet twenty-five years of age when he made known this, his great discovery, and which laid the foundation of his future world-renown.

In his famous work, the *Sponsalia Plantarium*, or the "floral nuptials," he says, "A great number of the plants lie with their stems concealed in the water, but when the time for propagation approaches they float to the surface with their flowers; thus *Nymphæe, Hydrocharis, Potamogeton, Persicaria Amphibia*. Others, again, are hidden in the water with all their parts, as *Myriophyllum, Stratiotes, Ranunculli*, and most of the *Potamogetenes*, but during their efflorescence they all show their spike of flowers, which when the seeding is effected again are pulled down."

Another curious and beautiful quality possessed by the aquatic plant *Vallisneria* is thus described. "*Vallisneria (Mich.)* possesses a rather long stem, but spirally contracted so that it appears very short. This grows in dykes and brooks under the water, and the stem has only one single flower. When blooming time comes the stem straightens itself until the calyx has reached above the surface of the water; when this has taken place the flower opens its chalice; but after a few days when it has bloomed and become pregnant she again is gradually drawn down in the water, by means of the stem again contracting itself spirally. This is the female. *Vallisnerioides* (*Mich.*) grows in the same places also under the water, but with a stem scarcely the length of a finger, and can thus not reach the surface of the water. This plant bears a great number of flowers which, when ready to bloom, liberate themselves from the stem, and float up like little blossoms. At first they are closed, but as soon as they have come up on the surface they develop and float about, and their pollen is now and again whiffed to the maidens floating about. These are the husbands of the former. *Michelius* noted this, and fully described it, but yet did not come to the conclusion that also amongst the flowers existed men and women."

After a short time, when the aged professor had procured himself permission to hold office vicariously, he appointed Linnæus to lecture in his stead on botany in the University. This was a very great honour, all the more as an older docens, Elias

Prytz, several years Linnæus' senior, had first been tried, but was found deficient in learning, and Linnæus had only been three years a student. Some of the wise men shook their heads, but the experiment was nevertheless ventured upon, and crowned with unsurpassed success.

With what heartfelt delight he now frequently lectured in the botanical garden to an auditory of some 400 students may be imagined. The eager listeners hung upon his words, for Carl Linnæus endowed the old neglected science with new life and interest through his vivid, lucid, and eloquent language. His youthful hearers became perfectly enthralled with the subject. Botany and Natural History now absorbed all interest from other branches of learning. Cicero's and Demosthenes's expounders were deserted to listen to him, and to learn more about the wonders of God's created world; dry discourses by professors had no longer any hold upon the eager students; the other lecturing halls stood empty—young, ardent genius carried the day.

Joy and delight glowed in every face of his

numerous disciples, when the throng of some hundreds of young botanists, under Linnæus's leadership, made for the fields and woods, intent upon botanical excursions. He divided his many followers into various groups to disperse in different directions in quest of rare and interesting plants, and when anything of peculiar note was found a bugle sounded which called all the students together to listen to a short and comprehensive exposition by their beloved young teacher. How this study in sylvan glades and meadows suited his ardent followers! They learned, as it were, to draw knowledge direct from Nature's own breast, and they became habituated to look on surrounding objects wherever they went, with other eyes than heretofore, and to perceive the order and harmony, the beauty and wondrous wisdom of everything created.

When the young men returned to Upsala from such botanical excursions, they marched in goodly order, joyously singing exultant patriotic songs, and decked with sylvan trophies, returning captors from a peaceful incursion in Flora's realm;

the bugles sounded triumphant notes, and while Linnæus hastened to the Botanical Garden, there to transplant the rarest spoils from field and forest, the happy students used to decorate the portals of his home with oaken leaves, or in his study erect a floral throne for their chosen monarch.

The news of Carl's success soon reached his parents in their rural home in the south of Sweden, and his mother was now no longer sorry that he had spent so much time in pasting dried flowers in his herbarium. She had thought he never would become anything more than an ambulance assistant, but both she and her husband now felt highly pleased to know that Carl, at only three-and-twenty years of age, discharged the duties of a Professor of Botany.

But the success of genius depends in a great measure upon the never tiring energy, diligence, and application, of its fortunate possessor, and like all successful great geniuses, Carl Linnæus was indefatigable. His days were devoted to his pupils, the number of whom were ever increasing, but the lone hours of the night he spent in his closet in conceiving and sketching the outlines for his great reform in Botany and Natural History. He has, himself, in his autographic notes, unostentatiously related how, during this period of his life, he wrote no less than thirteen important works in these sciences, and the which all he had already drafted before he was twenty-three years old; and of some of these he even made several copies himself. Everyone of these works became famous in after years, and were again and again considerably enlarged, until, with the subsequent works, great and minor, by Linnæus, all in all more than seventy, they formed a complete library in all the branches of Natural History.

With touching simplicity—for true genius is ever simple as its kindreds of truth and beauty and virtue—he wrote at this time, "the days of my life are short; what must be done, must be done quickly."

How thorough and diligent Linnæus was, even in his practical herborising, may be gathered from a short preface to a small treatise describing the trees and herbs that grew on a small isle in the famous lake Mälar, a water which is said to surround as many islands as there are days in the year. When Linnæus, at midsummer in 1731, went by sea from Upsala to Stockholm in a small packet belonging to his patron, Rudbeck, accompanied by between twenty and thirty students, they arrived after much rowing and pulling of the craft along the shores in a perfect calm at two o'clock in the night, at a small isle, where all the students and the crew lay down to sleep. But Linnæus, who had heard from the men that on this isle it was rumoured all kinds of plants and trees that grew in the kingdom were to be found, repaired instead up on land, and walked in a straight line up and down the whole length of the isle only leaving a couple of feet each time from the parallel line, "in the same way as a ploughman draws his furrows," so as to miss no piece of ground in his research; and he had barely thus completed his minute scrutiny, and just had time to gather a specimen of each plant, and some leaves of the trees, when the signal for embarkation sounded. But Linnæus had thus gathered eighty specimens of plants, and fourteen of shrubs and trees, nothing escaping him except the mosses, for which he had no time.

A splendid example this of thoroughness and diligence, well worth storing in the memories of those who wish to become successful in their pursuits.

CHAPTER IV.

T was but natural that such great success should evoke envy in less gifted and less fortunate teachers of the University, and Linnæus was destined to experience much bitter jealousy during his whole life—the unwelcome tribute which mediocrity always pays to genius; and as every country with more or less justice is famed for certain national traits of character, such as the English for "pluck," the Scots for "perseverance," the Finnish for "stubbornness," and so on, the Swedes are proverbial for "envy"—a national trait

of which Gustavus Vasa and Charles IX. complained, and from which every eminent Swede of both sexes has suffered much.

Nils Rosén, adjunctus in anatomy and botany, returned in 1731 to Upsala, from two years foreign travel, with senior claims to assume the vicariate of lecturing in these branches, instead of the aged professor. The result of this opposition in the first instance was that Linnæus resigned his situation as tutor to Rudbeck's sons. However, a new vista opened to him, for he had often heard spoken in Rudbeck's house of the journey which the professor's father, Olof Rudbeck, *senior*, had undertaken some thirty-seven years previously, to the northern Swedish province of Lapland. He had seen some of the interesting results in the fine drawings of birds before referred to, but which was almost all that remained, for the notes and preliminary work had been destroyed in a great fire which raged in Upsala in 1704, that town, as most Swedish towns to this day, being principally built of wood. The "Royal Society of Science" had again turned its attention to Lapland, which province, although not very distant, remained almost a *terra incognita* as regards its Flora, Fauna, and mineral wealth, far more so than such distant continents as Africa and America. Carl Linnæus, with great ardour, embraced the thought of a journey thither, and in a letter to the "Society of Science" importuned its president and members to realize the idea, assigning no less than eighteen various *raisons d'être* why such a journey should be undertaken; and finally, in all humility, offering himself as the most eligible explorer—though he modestly observes that he did not possess all the desirable qualifications. He was accepted, but he first determined to pay a short visit to Lund to perfect himself in the study of mineralogy, in which science he considered himself deficient, and which he thought his old friend Professor Stobæus's collection would enable him to do. However, in this he was mistaken, for he found on renewed inspection that it chiefly contained fossils. But Carl desired also to see and consult his parents before starting upon the perilous journey, so he visited dear old Stenbrohult in April, 1732.

What a meeting was that! How the old people looked with joy and pride upon their young, learned, handsome son. No cause now to rue the early love of flowers they had implanted in his heart and soul in his infancy. What delight to visit the old garden and his own old plot therein, still kept in trim by the loving hands of his younger sisters, who all seem to have inherited a love for botany. But the stay could only be short, and he soon set out again for Upsala. His mother wrote him shortly afterwards respecting his intended journey, which she feared would deprive them for ever of their new-blown hope; and amongst several reasons she used in trying to dissuade him was,

> "In thine own land live and dwell,
> Working there with humble faith,
> God will then provide thee well."

But his father left it to his son's own option. "You have only yourself to provide for," he wrote. "If you think it will lead to your future advancement, then pray that God may help you in this. He is omnipresent, even amongst the desolate Alps.

Put your trust in Him! My prayers to God shall follow you."

Friday, 12th May (old style), 1732, Linnæus set out on his journey to Lapland. The interesting diary of this his journey was translated into English and published in London, 1811, by J. E. Smith, the President of the Linnæan Society.

Linnæus wrote thus of the exact time of his starting. "At eleven o'clock, being at that time within half-a-day of twenty-five years of age. At this season Nature wore her most cheerful and delightful aspect, and Flora celebrated her nuptials with Phœbus." It is throughout a very vivid description; however, as we have an opportunity of presenting another version, not before published in English, taken down by Professor Roberg in his own quaint, peculiar style, on Linnæus's return to Upsala, it may have an additional charm of novelty even in the eyes of Linnæan students.

"1732, the 10th (more correct the 12th), May, I rode on horseback with a gun on my shoulder and extra-shod, from Upsala, with *Societatis lit de* recommendation *en general,* besides *Consistorii* to the places, also *Societ* money for the journey, and my instructions. Arrived the 12th (more correct the 14th), to Gefle, to Hudikswall, (Pastor Broman *Socius Societatis,* Rector Renmark) to Hernösand, to Skuulberg in Åugermanland the 20th, up which mount two men accompanied me, whom the sergeant gave me. We arrived at a robber's cave, up in the crag, and to a pole on which they were wont to tie their rope to travel up and down by; with great danger we got down again. In the time of Charles XI. there was one who had ventured up there to view it. The 24th I came to Umeå, was treated by the Governor of the Province, John Grundell. They were then tilling their lands, and were busy sowing their corn; of rye they have got nothing in their town. I then journeyed to Umeå Lapmark, although the Sub-Dean, Plantinus, dissuaded me in every way from my intent.

"In the Laplands no proper roads are cleared; paths cross each other; I had a countryman of mine, a peasant, with me from one to another. I arrived to the last hamlet in Westerbotten the 28th of May, past which flows the great Ume river: it is so wide that no one can shoot across it with a gun; flows calmly, and has its falls in other places.

"Should then proceed to Lykelde, the first church to be met with in the Lapmark, six miles sea-route (about forty English miles) by hop (boat) made by fir planks, thin and fastened together with ropes, equal in both ends; two can sit in it, sometimes also carry forty pounds dried fish, one of the men rows with two oars, the man carries it on his head one or two fjärdingsväg, (viz.: one = one and a-half English miles). The oars the man carries on his shoulders.

"I visited Pastor Gran, in Lüxela; there they also have a fair at times. I was treated to bread; themselves are very frugal; also butter, cheese, fish, meat, brandy and tobacco. There I got a man, a new settler in the clearings, who was exempted from all taxes, and had settled in the Lapmark where corn still can grow. We proceeded up the river along the strand, where there was no particular current, still we had to work hard at rowing; the ice was still lying along the strand,

and snow which had not yet began to melt, which else causes an unsurpassable current.

"The water rises up on shore, even into the forest. We came to the first Laplander in Umby, which is about ten (sixty-five English) miles long. Fir forest on morass land, pine where it is dry, herbs like here. We saw where the Laplander had had his cot of poles and fir-tree bark: I was tired, my comrade struck fire. We installed ourselves in the lap-cot; my comrade searched for the Laplander, and returned with him to me; the Lapp in his wadmal 'paita,' with his long staff, as tall as himself, knife in his belt, bag, needle-box, knife, rings to hang on, in the bag lies an *emplastrum*, enveloped in rind, a ditto small red earth, tobacco pipe, tin pipe, with inside of bone, they all smoke tobacco, a flat, round, bone spoon, steel and flint he buys at the fair, tinder he makes himself, the handle of the knife he makes himself, the blade he buys at the fair. All Lapps wear woollen shift or shirt. We seated ourselves three in the hop across a creek gave me food, cooked fresh fish—pike and perch, he gutted and cleaned the fish, he cooked, the wife mended clothes, the children ran slovenly about on the hill-side; everyone has his own dog, small kind, long furred. The food was put into a large wooden bowl, we eat with our fingers, no bread, the fish was partly sun-dried. My countryman, my former companion, bade farewell; with the Lapp I could not speak a word. I looked into the boat. Grass was beginning to appear by the strand the same kind as here. I was assisted forthwith by five Lapps more than twenty-five miles (160 English) Lüxela began to overflow more and more, so that it was impossible to advance further. We repaired to another river which coursed less rapidly. I came with my Lapp to a morass, one 'fjärdingsväg,' which we should pass. We waded in the water up to our waists; ice was laying under. He saw a tree, which he cut down and laid across a brook, standing on the one end, that I thus got across with my little gun on my shoulder, which I always carried with me, but we did not meet with our Laplander. We had to cross still another morass; I was wet and hungry. He struck a fire and I lay down, he left me to look for the Lapp;

I slept like a stock. I waited until the afternoon quite alone and quite at a loss what to do. This was the 4th of June. He returned, he had a small Lapp woman with him, and she a kettle with a pike, which he cooked, but no boat was to be had, and thus we had to return over the same morass; the woman left us. After some persuasion I bought of her a small cheese, made of reindeer's milk. We found our boat again, and descended by the stream; it went swiftly, I lay drying myself in the sunshine.

"The following day our boat went to pieces in the rapids; my stuffed birds and our alpine stocks drifted away, my book I had in my belt. The Lapps succeeded in getting to me on the rock. The Lapp waded to the shore—the axe was lost—he searched until he found a pole, in which he made a hole, the clothes were then tied to it, and those he first pulled across. After which I, naked, kept hold of the pole and followed after and thus came across, and tramped through the thick forest hungry and fatigued. I came to another Laplander and got fish, afterwards arrived at a settler's, where we grilled half-dried trout on the red-hot cinders—delicious, without bread; 14th June. Then came to my pastor in Lüxela once more, had then been days away and was happy to get food; then accompanied a peasant down to Granö, the 15th to Umeå, proceeded thence to Piteå, large high road, and took horse between the posting stations. The river at Piteå was impossible to navigate up the "Probst" rapids. I travelled to Luleo, arrived and visited the Dean Unærus. I there obtained a large boat, which four or five men pulled up the river, whom I had to pay as much as they demanded. Walked afterwards some three or four miles (twenty or twenty-five English miles), gathered herbs, stones, &c., left my bundles behind me, but now they are all here. I came to the silver melting furnace at Quikjok, stones I had gathered in the mountain, some seven miles distant (forty-five English). Now it is entirely deserted. A kind-hearted wife of a clergyman was there, her husband's name was Allstadius, she entertained me well. There I got a faithful interpreter and follower up amongst the Alps; my provisions were reindeer tongues, mutton, cheese, bread. We arrived in Halli-wari, always proceeding upwards, stones and earth, tall herbs, napellus, sonchus, &c.: V. Catal rubekij (*Rudbeckii Ol. fil.: Index plant, præcipuarum in itinere Laponico* (1995), *collectarum Upsal* (1720) *Actis hit. Suec, insertus,*) *curiosa saxifragia;* saw there mountains towering upon mountains covered with snow, no trees, no roads, no forest. We had also a Lapp with us, who had bought brandy from the parson's wife, carried it in a reindeer bladder. The Lapp was aggravated since I stopped now and then collecting herbs, which I found. We walked in shoes, else no one can proceed, made of ox-hide, colored red. The snow does not appear smooth, but rugged, like wavelets, with raised portions, caused by wind. The Alpine Lapps, kindly disposed people, their abodes grey baize on laths, where I stood under I could look out through the flue. There came home a herd of reindeer, some 700; a kind of gadfly greatly prevails among them. They milked them; the milk boiled, tasted like egg and milk, thick and nutritious.

"The Lapp took water in his mouth and washed the spoon with it; the hairs of the reindeer I skimmed off the milk, they strain it else through a hair sieve. I came to next Laplander with attendant. When I saw everything was similar at Piteå boundary I turned towards the Norwegian frontier.

"On the top of the Alpine range lies eternal snow, there I travelled for about seven miles (forty-five English) but hurt my elbow badly. The Lapps are afraid of clouds, it resembled a thick mist, was cold, drops fastened themselves to the clothes. We could find no road, were anxious and fearful lest we should freeze to death or fall down a precipice. We saw track of a reindeer, and gladly followed it up. The Lapp felt glad, I felt cold and fatigued, we travelled about thirty miles (200 English) into Norway; did not always find the Lapps cots and were glad to discover reindeer excrement. Lived upon fish, nothing but "rölingar," a kind of trout. We looked down into Norway, it is merely an *ora maritima*. We saw the sun above the horizon every night. It taxed us heavily to go down. We saw the western main, guessed that we were opposite Iceland.

"The Alpine plants disappeared, others were met with. Shortly before I felt cold in Lapp dress, but down here was so hot that I might have grizzled herrings on the flat stones, the grass was the height of a man. I lay down and eat wild strawberries and mesomora, I also found its variations. No wooden enclosures, the pigs are tethered with ropes, the tillage land sown with a mixture of oats, no rye, eat oatmeal bread. By the sea we saw ebb and flood, and many shells, *balani, stelleæ, marinæ*. I rested the 14th July, and lodged a couple of days by the sea-shore with a skipper from the "Norifjord," eat there cod-fish, smoked salmon. I accompanied a Lapp and ship's boat a bit out to sea. I came to pastor Rask, in Vörsta vicarage, a Norwegian parson who had been in Africa; the 16th saw his description in MSS. A hawker from Westgothland, was everywhere spoken about, he had told great yarns. I found that our clergy everywhere had wine.

"I again travelled over the Alps towards Torneå Lappmark, had no time to go farther into Kaitoma than forty miles (260 English) from the sea shore, and farther forty miles to the melting furnace, Kuikjok. In Kaitoma, the people fled as I approached; in the forest I found nothing different from what is here, but in the Alps, stones and herbs; tramping and food; but to keep an interpreter was too expensive.

"At the furnace I rested; I suffered from indigestion from having to eat too much cheese.

"The tarns on the Alps are from stagnant water, milk white, somewhat reddish, turns into shiffer. From Kuikjok to Luleå travelled twice on rafts, could get no boat. By the pearfidhey at Purkijaur. Journeyed four miles towards East-botten; could not speak Finnish, and Swedes there were none; compelled to return; came to Calix. The district judge Höijer intended to travel beyond Torneå. Suanberg promised to show me the art of probing. I remained there eight days. Journeyed up to Kengis or Sappawari copper factory, twenty-seven miles from Torneå, in the month of August; then to Joneswando, seventy-five miles beyond Torneå ironfoundry; then down to Torneå. Travelled on the other side to the posting-station; polite conference with the county constable, an important personage. I got a horse; arrived at Uleå; treated by the dean of the church in Uleå.

"Finland, Brahestad, Old Carleby, Jacobstad, New Carleby, Wasa; visited the Burgomaster, kind man; thence to Christina, then to Björneborg; dangerous, stony, horrible road. Passed Nystad. Arrived at Åbo the 30th September; five days on seven miles (forty-five English) highroad, after which passed by little boats the gulf of Bothnia to Grissleham, the 8th October, to Upsala on 10th; travelled 650 miles (4,200 English).

"The hill of Norrby lies four miles from Hudikswall. Skullberg by Hernösand in Ångermanland. Åsila Lapmark is Ångermanland's Lapmark.

"In Umeå the governor of the province has potatoes in his garden.

"The river Juktan, twenty-five miles above Lyksele. The snipes make a noise as if they were laughing. The Lapp calls his boat "håpen." A reindeer milks barely a quarter of a tumbler full. "Fir and pine in Lapland are all turned the same way." Linnæus wrote a letter to the same professor Lars Roberg while on his Lapland tour, from which the following is an extract; also, we believe, rendered for the first time in English:—

"It is an old custom to send anybody who won't turn to any good on board a craft, but I know a better way, viz., send him who is very troublesome up the Alps. Never would I have entered upon such a journey—a way so full of innumerable and evident death-gaps—if I could have foreseen it. No. *Crede mihi bene, qui bene latuit, bene vixit.* At any rate, old St. Paul's saying comes true, also as regards me, 'I have travelled far.' I have already journeyed for more than 300 miles (2,000 English); yea, if I count all, nearer 400, since the last time I had the honour to speak with you. I have seen solemn inoccidum in the coldest winter. I have felt so cold in the depth of winter that fingers and face might have been frozen off, although the summer was in view, where it was so hot that turnips might have been broiled against the mountains. I have been precipitated down mountains half a *fjärdingsväg* (seven-eights of an English mile) at one fall, and yet was saved. I have been down hollows where the water had undermined the snow, and they had to haul me up with ropes. I have been a target for the Lapps on the Norwegian side, who use not to miss their mark. I have been in danger at sea on the western main. I have been in perils in rivers, &c. Curious has my journey been, but God help him, who must pay as dearly as I have for my curiosity. Stones and minerals I have got a great collection of; many birds, some insects and fishes; a heap of *conchilles* and *zoophyta*, and a great many herbs.

"When I ascended the Alps I might not have known whether I was in East or West India, so entirely was the world around me changed; so many strange objects appeared. I saw nothing there, but naked mountains surmount each other; no forests, no trees, no houses, no wooden inclosures, no roads, no singing birds, no setting of the sun; few herbs of what I had formerly seen were here found; all was new, all strange. *Ovidii descripto ætatis aureæ* (if only the snowy Alps be excepted) seemed to fit *quadrera*. I was so long amongst the Alps that I thought I could accustom myself to the Laplanders mode of living, the language excepted. Their 'miissung,' which I had morning, noon and night, I got at last so tired of, that I, with the murmuring children of Israel, wished for meat, yea, a morsel of bread. The Norwegian people did not receive me very favourably, therefore I did not remain with them very long. Now I intend, God willing, to proceed to Åbo, and from thence to Upsala, to which God Almighty help me. I am already surfeited of so much travelling I am quite done up by too much running with the Lapps in the Alps, with whom I battled together more than 100 miles (650 English).

"I am tired even of writing, and reserve the rest for oral communication, and beg you, sir, my honoured professor will graciously accept of what I write in haste."

Fifteen years after this, (in 1747,) Linnæus wrote a *pro memoria* to the "Royal Academy of Science" in Stockholm in reference to this his early travel in Lapland, the Academy having solicited his opinion upon the province in question, (presumably now translated for the first time into English).

"The Royal Academy of Science has communicated to me, as being their member, an extract from the protocol of 17th October, 1747, by the Royal

Directors over the Ecclesiastical Department of Lapland, thereby to learn my candid opinion.

"I am not a little pleased the Royal Directors vouchsafe to turn their eyes to the cultivation of the Laplands, since they constitute a considerable portion of Sweden, if not half of all Scandinavia. We look with astonishment at what manufactories have accomplished in the southern countries of Europe, but here we find a thousand times more of what nature displays before the eyes, although the most of it until our times has remained unknown. To get this explained, there is no other means than by the Theologos who will be residing there.

"They could teach us the *height of the most towering Alps* towards the western main, and of which those opposite Torrtjord are so high that, as regards myself, I doubt if any surpass them in Europe.

"They could find out if the *snow on the mountains yearly increases*, and from its crevices calculate how many years and centuries it has lain.

"They would make attempts with thermometers through the Lapps to find out what is *the severest*

The Floral King.

cold on the Alps, when these catch the wild Alpine snipes, and they could tell us how great *the heat* becomes in the summer between the Alps.

"They could discover various useful kinds of *stones* amongst the crags, which consists of different kinds from ours.

"They might teach people to gather and hew these stones (*Saxum tritorium*) from which the Norwegians make those splendid ' *murkqvarnar* ' (millstones) which they sell all over Europe.

"They would find out where *the iron sand* is generated which by the rivers is brought to *Sinum bothnicum*.

"They would be sure to discover *tin-metal* in the Alps, where all stones are full of granites; yea, infallibly also of *gold ore*, besides other precious metals; if they themselves understood the stone sorts.

"They could seek to find if in the Alps were any chalk with *petrifications*, and if there are any proofs of the shores of former seas.

"They would be sure to find *umbra* and *other rare earth sorts*, all the more as the earth there is very brown.

"The short time I passed there rendered me more than a hundred kinds of plants, which were unknown to the greatest botanists. I doubt not but that several hundred more will be found by those who here would have opportunity to search through hills and valleys, and thus be the means of increasing our Natural History, to the admiration of the learned world.

"I have been able to get plants, *Caput bonæ spei*, from Japan, from Peru and the Brazils to our Swedish Hortus Academicus, but as yet not one from our own Alps, although I am a Swede myself. If I could obtain them, so as to propagate them, I could easily for the superfluous get in exchange Palms, *Musæ*, *Radix Ninsi* and all the rarest plants from the choicest gardens in Europe.

"*Radix Archangelicæ, Radix Gentianæ, Radix Rhodiæ*, grow on our own Alps, are then gathered by the Norwegians, sold to the Dutch, then to the Germans and by the Germans to the Swedes; why then could not our parsons as well sell them to us, and thereby make some money? If any one of them would make a plantation of *Gentiana*, I am sure he would better himself and serve our country. If the herbs which grow on foreign Alps, and yearly are imported by our druggists, such as *Victorialis, Spica celtica, Crocus, Daucus creticus, Bistorta, Helleborus niger, Doronicum, Carlina,* &c., with satisfaction are to be planted, I hold that Lapland is the place. In Lapland Theologians would have the best opportunity for describing the many birds which gather here during the summer from the whole world; and here to utilize the aquatic birds and their feathers, which so greatly benefit the foreigner but leave us nothing but the mere beholding, while they here lay their eggs and bring up their young. Of the natural history of the *Filfras* we are more ignorant than of that of the Paradise bird, the Hermilines, the Alpine mice, dogs and snipes, and several Lapland arrivals' traits are still shrouded in a dark mist. Rauder has never yet been drawn, and much more which belongs to the Fauna. We here find a people who live only on *animals* and *fish, without salt, without vegetables;* thus, in such a cold climate, they must have their peculiar maladies and their own cures. It ought to be ascertained from whence accrues the violent colic with blood discharge, their frequent headaches, their lumbago, and their weak eyes, &c. It ought to be computated what is the average longevity and what sickness carries off the majority, when first proper remedies could be prescribed.

"Here ought to be ascertained what kind of contagious diseases some years, like the plague, carry off the reindeers, that a remedy might be found, for they are the foundation of economy, life, and well-being of Lapland.

"Here much might be enumerated, which I consider unnecessary to do, because all these defects arise from the same cause and can all be remedied by the same means, viz: attention. As impossible as it is to read a book without knowing the letters, so impossible it is to solve the mysteries of nature without knowledge of its objects. I have endeavoured, from the first day I came to the University, to teach the students the knowledge of stones, ores, different herbs and trees, animals, birds, fishes, and worms, diet, and various kinds of diseases; I have made it briefly and systematically, so that it would not burden the memory nor require long time. Many hundreds have in these respects made considerable progress, but yet, with the exception of two, who last summer lent me their ears, no one of all those who were intended to be clergymen in Lapland have ever heard me, although I have offered them all, at examination, to attend my lectures without the slightest remuneration. I am sure that these [except that they with becoming diligence study theology and the Lapland tongue, that they instead of devoting themselves to a number of philosophical sciences, which are less important in Lapland,] were obliged to devote some application to *historia naturali*, all this would be gained quite of its own accord, without which knowledge all good intent is impotent. When new *Ecclesia* have been properly planted in Lapland and when afterwards students who intend to go thither have become more mature at the University, it would be no small assistance if proficiency in natural history, *cæteris paribus*, would give them some claim to their appointment. By this they would find themselves working for a useful science, and to themselves equally beneficial. They would then no doubt vie with each other after having come to the Laplands, in gathering tests, make experiments to send in to the Royal Academy of Science, by that to serve the public and their Lapland, when now often they do not understand more than peasants what the Creator of Nature has so gloriously placed before their eyes.

"These are briefly my simple thoughts, which the Royal Academy of Science, with the other members, may vouchsafe to send in to the Royal Directors, who perhaps from this may find a reason to benefit Lapland and in that our country."

During his Lapland tour, Linnæus saw for the first time the sea wheat-grass, the same which grows in Iceland and in the Faro Islands of Scotland, where its spikes are gathered and prepared for bread, he thought it suited for Lapland, but although it is a perennial plant, and mostly cultivated on sea-coast land, he never succeeded in persuading the government to interest itself in its behalf by introducing it in these sterile regions, and Linnæus maintained

The Floral King. 65

that unless the clergy were made to set the example by using it in their domestic economy the prejudices of the Laplanders and the settlers would not be overcome. And thus it is scarcely ever heard of in Lapland to this day.

Amongst the few people of note Linnæus met with in the Lapland settlements deserve to be mentioned Pehr Fjällström, schoolmaster, and afterwards pastor in the parish of Lycksele, for he had published a Lapp grammar and dictionary, and translated into that tongue a primer for children, the catechism and ritual-book, and the New Testament, and collected a Lapp hymn-book, which went through several editions.

It may be interesting to cite that four years after Linnæus' Lapland tour, in 1736, Professor Anders Celcius, together with Messrs. Maupertius, Clairmont, Monnier, and Canns, repaired to Lapland to take scientific measurements by which they, after six months labour, succeeded in verifying Huygen's and Newton's well founded opinion that the earth is flat at the poles.

CHAPTER V.

URING Linnæus' journey in Lapland he stopped for some days at Calix, and there acquired the art of probing. On his return to Upsala he began lecturing on this as well as Botany. But an intrigue was afoot amongst his colleagues, stirred up by the mortification they had felt in seeing their own lectures neglected by the students in preference to those of Linnæus, and they prevailed upon his rival and antagonist Adjunctus Rosén, who was recently married to the niece of the Archbishop, to exert his influence, and an injunction was published

by the Chancellor of the University that no Docens in Medicine should be accepted, and that no one was to be permitted to hold public lectures who had not themselves passed public examination for this purpose—a paragraph which had been overruled in the case of Linnæus, when he had been accepted as Vicarious for Rudbeck, where others had failed. Linnæus mainly depended upon this, and he saw at once his means of existence, as well as his immediate future hopes ruthlessly crushed by an envious faction, and his adversaries had counted upon this with villainous foresight. Adjunctus Rosén lectured in Anatomy and Natural History, which two branches also belonged to Professor Rudbeck, and he tried even to get that of Botany, and this Linnæus declared he would give up, but which Rudbeck would not allow. Rosén even took private instruction from Linnæus in Natural History, as he was inferior to him in this respect, and yet he tried to oust him in this very branch, and succeeded also. By threats he prevailed upon Linnæus to lend him some of his valuable manuscripts, which were the most precious things Linnæus possessed, and he clandestinely copied them, but when the ill-used author discovered this, neither threats, nor guile, could prevail upon him to allow his unprincipled disciple and opponent the use of the remainder.

Linnæus was of hasty and somewhat choleric temperament, easily roused to joy or sorrow, but the fire of his wrath soon passed away, and he was immediately ready to forgive. Such characters are the most amiable, but they are sometimes in the heat of the moment transported beyond themselves, and the electric fire has flashed with anger before calm reason has had time to stay the impulse. Although he had mastered the knowledge of wondrous Nature's wisdom, yet he was a perfect novice in that of the world. He understood not to meet man with his own weapon, deceit, by seemingly conforming to his adversaries' wishes, to gain the end in view, for dissimulation was an art entirely unknown to the young sage, with a heart as pure and simple as that of a child.

And yet, withal, ambition was his ruling passion, that noble ambition which seizes great souls, and carries them on daring wings to accomplish feats of which lesser minds cannot even conceive the unlimited benefits gained by the victory, although they may daily enjoy the results.

Linnæus had been wounded in his most vulnerable point, they had tried to stifle his ambition. Who can wonder then, that the fiery young man, with drawn sword, rushed in to his cowardly foe, and demanded reparation for his wrongs. Fortunately for all, and for Science, Linnæus was stayed at this culminating point by the timely interference of a friend, who threw himself between the antagonists; and Linnæus's judgment soon got the better of him, and he left the terrified Rosén to the admonition of his own conscience, and that Rosén, though late, repented, after-years proved.

The high-minded young botanist repaired to his lonely chambers, and made a vow never again to seek revenge upon his enemies, but leave that to be dealt by Him who watches over the destinies of all. And he put over his door this motto: "Live irreproachably, God is ever present," a maxim to which Linnæus proved true the whole of his remaining long and illustrious life.

This incident has lately been refuted as unfounded by the learned researcher amongst Linnæus's unpublished letters, but until this most painstaking Linnæn student chooses to publish these facts he promises, the result of some sixteen years' studies of the Linnæn literature, we must give credence to the two chief biographers of Linnæus—Swen Hedin and Stöver, of whom the former was a pupil of Linnæus, and stood in near relation to him at the University of Upsala, and he has with much detail related the challenge alluded to. The affair was duly reported to the Univerity authorities, and it was resolved that Linnæus should be expelled the seat of learning. A man destined to become the greatest honour to the University, and who has brought more glory upon Sweden—the reflex of his learning and genius—than any other sage or hero. This Heaven-gifted, simple-minded Carl Linnæus received such a bad certificate for acquirements from the Academy, that he could not show it to procure himself admittance to the University, and now he was to be expelled the University for bad conduct. Strangely turn the destinies of men. We must sincerely and humbly thank God that such a thing as ambition exists to lift strong souls on gigantic wings above the common herd, that they may soar away to loftier goals beyond the ken of envy and malice.

But Dean Celcius, who saved Linnæus from breaking his career for poverty's sake, stepped in once more, and warded off the threatening blow. The air of Upsala, (that ever the sacred groves of Odin, and Freya, and Balder should be polluted by envy, and rancour, and cowardly intrigue,) was no longer genial to Carl Linnæus, he therefore eagerly embraced an opportunity of continuing a scientific research amongst the mines and melting furnaces of the neighbouring province of Dalarna or Dalcarlia, from the Governor of which, Baron Reuterholm, a letter arrived requesting Linnæus to continue the peregrination which he had begun in that district the previous summer. Our young sage took with him seven young and eager students to benefit by their common toil. Linnæus writes of himself from this period, that "during the days he crawled amongst the stones in the mines, and the nights he passed before the fires of the smelting furnaces."

Falun, the provincial capital of Dalarna, was the residence of the Governor of the province, who also in his turn became charmed by Linnæus's brilliant genius; as well as his house-chaplain, J. Brovallius, who rose to become Bishop of Åbo in Finland, through whose exertions the Bible was translated into Finnish, 1758. The sons of Baron Reuterholm became Linnæus's eager disciples in the much frequented lectures he delivered in the art of probing and mineralogy. Linnæus also practised as a Physician in Falun, and was much consulted, so that he earned and even saved some money. The best families in the little town vied to see him as guest, and flippant society, for once, during a whole season, gave up its petty diversions to listen to the lively and enthralling descriptions of natural objects by the gifted doctor, who soon found himself the centre of attraction and the soul of every gathering. Linnæus, admired by all, had then good reason for writing about those days, "Everything prospered here with me;" and here he was to cull the softest and most ethereal flower that grows in the garden of man's life, his first and only love.

In Falun lived the district Physician, Dr. Johan Moræus, at whose house Linnæus was a frequent guest, and with whose charming eldest daughter, Sarah Elizabeth, our young hero fell deeply in love. That the tender passion was returned he felt certain, through the responsive ardent glances of her beautiful eyes, although he had not as yet ventured upon declaring himself, for he had at present too poor a prospect to offer the daughter of the learned and highly respectable doctor. However, his intimate clerical friend, Brovallius, advised him to do this, the sooner the better, for as the young lady's father was considered a rich man, the pecuniary prospects of Linnæus would also be better through the engagement, and it was time that Linnæus should proceed to Holland to take his medical degree, as was then the custom, and, after a stay of some three years, he would return to Sweden, with an assured position, as a physician. Carl sought and found the opportunity of ascertaining from the lips of his beloved that their passion was really reciprocal, and on speaking to her father, Linnæus, to his great surprise, was accepted as his future son-in-law, provided he first repaired to secure the medical honours in Holland. With but slender means he, in 1735, started on his long journey, having obtained a small scholarship from Upsala: and on parting, his young betrothed pressed into his hands a small purse worked by her own hands, and filled with ducats, for of course she longed to help the object of her love. Carl had written a few verses to her on leaving, in which he addressed her as his " Friend Moræa," for it was usual in those days that the name of ladies' surnames were ended by the feminine a, thus Moræus's daughter was called Moræa, and Linnæus's future wife would be called Linnæa; and it is a curious coincidence that thus his lady-love and the flower he loved the most, and which also adorned his seal, his portraits, and his future escutcheon, were both called up after him, Linnæa, and that this charming child of the Flora of the North, with its pale carnation colour and delicious fragrance, invariably grew two upon one stem, a loving couple enjoying a floral sylvan life together. Carl, in his journey south, visited his home at Stenbrohult, where everything was as of yore, with the exception that his loving mother was dead and his father much aged of late. The old man grieved that he could not assist his son with any money for his foreign sojourn, for few students ever ventured to Holland to take their degree with less pecuniary means to rely upon for sustenance; but Carl Linnæus was, nevertheless, full of hope and confidence. He carried very little of the magic gold in his knapsack, but he had a mine of wealth in his learning, and magic in his genius, and he took with him the manuscript draughts for many great works, the which he had conceived during his three years in Upsala. He has himself given us a list of them, the length and importance of which certainly is most astounding, and naturally forces the thought upon us, how few, even of the gifted sons of earth, have ever done the like in so short a space of time.

Linnæus wrote himself regarding these :—

"As fruits of my studies I can show the following works in manuscript by me, *Propria Minerva* elaborated, viz:—

"1° *Bibliotheca Botanica*, which criticizes all books arranged in order, all *Methodicorum Assectæ* under their *Primores*.

"2° *Systemata Botanica*, when all *Botanicorum Theoria* are shown in *compendio*.

"3° *Philosophia Bontanica*, whereas all *Botanici* have not had more than twenty or thirty general bases, I have here brought them to 200 or 400.

"It shows first how male and female propagate amongst the plants in almost the same way as amongst animals, whereby one ought to know all plants by the first look at them, and in which *Botanici* have been mistaken when they have made the new, wrong systems.

"4° *Harmonia Botanica* shows how the names of all plants ought to be made, and that not the tenth part of *nomina generica* is proper, that no *nomen specificum* is correctly made how these ought to be formed.

"5° *Characteres Generici*. To know all plants at the first glance by the definition of the flower is here practically proved, and that these *characteres* can be applied to all methods, which no *Botanicus* has understood before.

"6° *Species Plantarum* under their *genera tomi duo*.

"7° *Nuptiæ Plantarum*. In Sweden no method has yet been made, but abroad generally one in each land; to invent a general one, is the most difficult in Botany, therefore I have tried one on a new *principio*, when all others are spurious, mine is not. This book is now in Germany to be printed.

"8° *Adonis Uplandicus*, or garden herbs in Upland, described by the students, also sent abroad to be printed.

"9° *Flora Lapponica* describes the herbs, shrubs, and trees, which grow in Lapland, so embracing that also all *Fungi* and *Musci* are observed, together with the use and usefulness of each with the Lapps, besides figures and descriptions of more than 100 rare herbs, most of which have never before been seen, much less ever described.

"10° *Lachesis Lapponica*, a handbook of

Lapland's appearance, husbandry, apparels, chase, &c., in Swedish.

"11° *Aves Sveciæ* more than 300 species of birds described, observed in *Sweden*, and one learns to know them at the first glance.

"12° *Insecta Uplandica* here are described 1,200 insects, observed in Upland, gathered by me, and still preserved.

"13° *Diæta Naturalis*, teaches *ex principiis Zoologicis* in a hitherto unknown manner, how a man can attain to great age, affirmed by people who have lived long, and how a feeble body can be sustained a long time.

"Upsala, 1st October, 1733,

"(Carolus Linnæus)."

CHAPTER VI.

INNÆUS proceeded from Helsingborg, in Sweden, to Germany, and after a short stay at Hamburg, continued his journey to Hardewijk, in Holland. He published here a treatise full of genius, and in June took his degree as Doctor of Medicine. He then left for Leyden, at which famous University he became acquainted with many distinguished and learned men. Joh. Fred. Gronovius, Senator in Leyden, interested himself much for Linnæus, and as his little money was already spent in travelling, Gronovius caused

Linnæus' book, "*Systemæ Naturæ,*" to be printed and published at his own expense. This, his great work formed quite an epoch in Natural History, and gradually, during successive editions, became, by Linnæus, so much enlarged, that the twelfth and last edition published in Stockholm, 1766-68, reached 2,300 pages. Many different translations and extracts and adaptations have, from time to time, been published in many lands, amongst others, one in Batavia, between 1770-80, by a society which had translated the nomenclature into the Malay tongue.

Gronovius himself was, at this period, engaged in determining and describing the herbs and plants which J. Clayton had gathered in Virginia, which now, with the assistance of Linnæus, were arranged according to his sexual system. Linnæus and van Rogen also became friends.

Also with John Lawson, a learned Scotsman, Linnæus became acquainted, probably the author of "History of Carolina," containing the exact description and natural history of that country, London, 1709-19, and all these learned gentlemen urged the young Swedish doctor to publish his manuscripts. But Linnæus tried long in vain to see the famous physician, chemist, and botanist, Dr. Herman Boerhaave, for this celebrated man was constantly besieged by visitors, so that many great people even had to wait for hours in his antechamber before they could be admitted.

A story is extant that even Czar Peter, [we suppose, however, it was when he worked incognito as a ship carpenter at Zaardam] had been kept waiting two hours for an audience, for famous though Boerhaave was, it is a puerile boast to say he kept the Muscovite Czar waiting, had he known who he was, and much less credible it is that impetuous Peter would have brooked such an insult, if it had been the learned and famous physician's desire to make royalty for once wait and dance attendance upon science. Another anecdote tells of a Chinese Mandarin having written a letter to him only addressed " Herr Boerhaave, famous Physician in Europe." Linnæus attended during eight days in hopes to get an audience, and at last succeeded in making himself known, which lately has been ascertained was three weeks before his famous "*Systemæ Naturæ*" was published, but which else has been considered the "open sesame," that gained access to the famous Hollander. Boerhaave sent his own carriage to Linnæus' distant lodging, and by letter invited him to the country seat of Boerhaave, where the latter with impatience was waiting for the arrival of Linnæus. On arriving there Linnæus was exceedingly well received by the host, who introduced him to many of his learned friends, who were there to meet him. Boerhaave possessed a great and rare collection in Natural History, and his garden boasted the choicest and most curious exotic plants. On shewing these, he found great pleasure in putting the learning of the Swedish doctor to the test, and the day passed in agreeable conversation, erudite researches, and even contentions, which only served all the more to cement the friendship between them. Boerhaave became so interested in his young guest that he was loath to let him depart, and made him the most advantageous offers if he would remain in Leyden. But Linnæus had decided to go to Amsterdam, and could not be persuaded to remain. Boerhaave then gave him a letter of introduction to Burmannus, who was Professor of Botany in Amsterdam.

When Linnæus came to Burmannus, he found him intent upon arranging a collection of natural objects, which had just arrived from Ceylon, and he stood perplexed how to decide to which species and order belonged a tree of which he now, for the first time, saw the leaves and blossoms. Reading in the letter from Boerhaave, that the deep insight of Linnæus could not be sufficiently extolled, he did not give himself time to greet him before he first eagerly put the difficult questions regarding the strange tree, and Linnæus gave him, there and then, a highly satisfactory definition. Burmannus received him as his honoured guest for the time he choose to remain in Amsterdam, which Linnæus gladly accepted, and this procured him admission to the garden for the cultivation of medicinal herbs. He opened his collections for him and told him of his own experiences, and introduced him into the best and most learned society, which procured Linnæus many patrons. Linnæus assisted Burmannus with his work about the natural history of Ceylon, which as "*Thesarus Zeylanica*," was published in 1737 in Amsterdam.

During this, in every respect, happy period, Linnæus published several of the treatises he had already written at home, and taken with him. The winter of 1735, was thus passed with alternate work and great enjoyments, when one day, to Linnæus' great surprise, he received a visit from the burgomaster of Amsterdam, a rich banker, George Clifford. This gentleman had stored his large garden at his country seat, Hartcamp, situated between Leyden and Haarlem, with the most choice exotic flowers from many climes, for which he had excellent opportunity, being a director of the Dutch East India Company. But his great collections lacked the systematic order of science, this he felt himself, and his friend Boerhaave had also given him to understand that, and advised him to invite Linnæus to arrange them according to his system. For he said, "this Swedish doctor is the greatest botanical genius we have." Clifford came to invite Linnæus and Burmannus to Hartcamp, thither they accompanied him, and when there he told Linnæus that he wished to offer him the place of house-physician, and manager of the grand and beautiful gardens, which at once captivated the ardent horticulturist. This was an offer not to be despised, and although Burmannus was unwilling to allow Linnæus to leave him, still he was obliged to renounce his prior claim, for the great advantages this brought his guest. Linnæus soon removed to Hartcamp, where he lived in the Chateau, with numerous servants at his command, and a carriage and team of four horses for his disposal, whenever he drove to visit Amsterdam. He was permitted to send for all plants he wished to enrich the garden with, and to buy what books he considered the library ought to possess, and had thus a splendid field in which to exercise his genius. One day, during this happy time, Linnæus was surprised by the visit of an old University friend from Upsala, Artedi. He had left Sweden a year before Linnæus, to study Icthyology, the science to which he particularly devoted himself, and for which purpose he had repaired to England, where he had been well received by Francis

Willoughby, and in particular by Sir Hans Sloane, who had afforded him every facility for his study of fishes. Now he had to come to Holland, so celebrated in those days for learning, to further pursue his science, and take his degree, but he was in quite a destitute condition, and Linnæus, who had himself experienced the stress of poverty, afforded him every means of assistance, and the two young learned men agreed that should death overtake the one, or the other, the survivor should save the manuscripts and belongings of the deceased, and consider it a sacred duty to see the learned researches published.

Albertus Seba, an aged apothecary in Amsterdam, had published one volume upon quadrupeds, another upon serpents, and now desired a competent coadjutor to assist him with one about fishes, as his impaired health prevented him from assiduously applying himself to the task. He had solicited Linnæus to help him, but he could not spare time, and Linnæus therefore hastened instead to recommend his friend Artedi, as the very man for the task. He was accepted, and proceeded with the work, but one night, when returning late from his hospitable employer, on crossing one of the many canals of the Dutch town, he mistook his way in the darkness, that prevailed, and fell into the water, where he found his early grave, in that element the inhabitants of which had been the objects of his scientific researches. Two days afterwards Linnæus learned the sad fate of his much-loved friend, and hastened to secure the manuscripts, regarding which the two had had such strange presentiment, but Artedi's landlord had sequestered them for debt. In this emergency poor Linnæus, who knew that money could not for very long, if at all, be obtained from Sweden to redeem them, had no other resource but to turn to his patron; and noble Clifford, without a murmur, paid the required money, and the valuable ichthyological manuscripts were saved, and afterwards, under Linnæus' auspices, published in Leyden, 1738, by which Linnæus secured his poor friend Artedi's name in this particular branch of science.

For Artedi's works, which required to be clearly copied, Linnæus employed a needy young Swede, come to take his degree in Leyden, and whose pecuniary means likewise were exhausted, a very usual failing with travelling Swedes it seems, quite another national failing; by this Tiburtius Kjellman Artedi's work, at last, was made ready for the press, and his name handed to posterity.

❈ CHAPTER VII. ❈

N 1736 Linnæus went to England, the expenses of the journey being defrayed by Clifford, as one of the principal reasons for going was to acquire from the Apothecaries Garden, in Chelsea, those plants which the gardens of Hartcamp did not possess. Linnæus was furnished with a letter of introduction to Sir Hans Sloane from Boerhaave, but as it was couched in highly eulogistic terms, it had a contrary effect to that intended upon Sir Hans, so characteristic of a true Englishman, and Linnæus was therefore

rather coldly received. However, Sir Hans soon found that the letter-writer had not exaggerated, and Linnæus was favoured with his confidence and sincere respect. Linnæus wrote, on his return to Holland, in letters to Sweden—" With the English I agreed very well," and also, " Sloane's great collections are quite in disorder."

But when Linnæus in *Hort Cliff* records the species Sloane, he remarks, " This name holds its place amongst the illustrious. To Hans Sloane, President of the Royal British Society, is owing nearly all our knowledge of Japanese also of many American plants. He alone collected many things, more than any one else in Natural History; arranged them in a Museum of which the like does not, and scarcely can, exist. I boast that I have seen in such, a Herbarium of Sloane, Plukenetus, Petiverus, Camellus, and of very many other celebrated Botanists of the past."

Philip Miller, the Manager of the Chelsea Medicinal Garden, did not receive him very well, either at first, and even openly ridiculed him, saying, " A nice man truly, this Mynheer Clifford's

Botanist, he does [not know a single plant;" but as he gradually became acquainted with Linnæus, and his new system, Miller changed his opinion, and became instead a great admirer of Linnæus's genius, and allowed him to take what plants he desired from Chelsea to Hartcamp; those which he selected were principally American. Miller had just published in London, *The Gardener's Dictionary*, which afterwards went through nineteen editions, and was translated into many foreign languages. Nearly twenty years afterwards (1755) he also published 2 vols., *Figures of the most beautiful, useful, and uncommon plants, described in the Gardener's Dictionary; with* 300 *plates*." He died an octagenarian in 1771.

Another English Botanist and Zoologist, contemporaneous with Linnæus's visit to England, was Marcus Catesby, who had travelled in Virginia 1712-19, and in Carolina, Georgia, Florida and to the Bahama Islands 1722-26, the latter journey at the expense of William Sherard and Sir Hans Sloane. He had (1731), in 2 vols., published the results of these his travels, under title, *The Natural*

The Floral King.

History of Carolina, Florida and the Bahama Islands, containing the figures of birds, beasts, fishes, serpents, insects and plants; 200 plates, with text in English and French." Linnæus's opinion, however, was that " Catesby was not very particular, with the exception of his drawings, and he was himself not proficient in Natural History." Linnæus desired, in particular, to visit the celebrated Johan Jacob Dillenius, a native of Darmstadt in Germany, but now Professor in Oxford, who had a great renown as a Botanist, having for eleven years been the Manager of the famous Botanical Garden at Eltham in Kent, owned by the Brothers Jacob and William Sherard, the latter of whom had for many years been Consul at Smyrna, where he had begun his great collection of plants. On the death of William Sherard, in 1728, Dillenius was appointed to the Chair of Botany in Oxford, for which professorship William Sherard had left the pecuniary means, with a view that his large collections might be scientifically arranged and described, to form a supplement to Casp. Bauhinus's *K.I.U.A.E. Theatri Botanici*, Basel, 1671.

Dillenius agreed to perform this work, but from various causes was prevented from effecting it, and no visible fruit of his labour was left at his death. Dillenius had, however, amongst other things, published *Hortus Elthamensis*, which became very famous, but his principal work was *Historia Muscorum*, Oxford, 1741, 4 vols., with 85 plates, in which 600 species were described, and is considered a cleverer work, that has handed this author's name to posterity. However, neither did Dillenius receive Linnæus well at first, and said of him, "This is the young man who brings confusion into Botany." But also he changed his opinion, and became a most ardent supporter of Linnæus's system. He prevailed upon Linnæus, at last, to remain a whole month with him, and the two friends were scarcely ever apart during the day; and when, finally, Linnæus was compelled to leave, Dellenius said, with tears in his eyes, "Dear friend, remain with me; let us live and die together; my salary is sufficient for us both.

Linnæus himself, in a letter of this period to his

venerable friend Professor Olof Celsius, senior, at Upsala, writes, "My *Genera Plantarum* have at last, this week, been struck off at the printers. Dr. Dillenius got half of it when it was printed, as he desired. When he saw how I had treated his *Genera*, that gentleman became as angry with me as before he had been affable. When I came to Oxford he would scarcely ask me to come in with him; at last he exploded with slight innuendoes and contemptuous looks. I was for three days with him in the town, and was scarcely allowed to see a plant. I paid the carriage in his presence, by which the following day I was to depart, as he meant *mir wohl*. I could then no longer bear it; began resuming old arguments, when we proceeded to examine *flores*, and allow the autopsy to judge between us; then at last we agreed. I was then forced to leave my travelling companion, and, *nolens volens*, remain. Ever afterwards we were scarcely parted from one another two hours all the time I remained in Oxford, and when at last I departed he let me go with tears. He presented me with his *Hortus Elthamensis*.

"Sherard's collection is in *Europis sine pari*, but in exotics rather musty.

"*Hortus Oxoniensis* is a good *plantis Europæis*, but hybernacular and hypocasta are rather empty.

"Now he had began a clear copy for the last time of *Pinacem Sherardi*, but had not got far. I wish his *Historia Muscorum* could be published, it is excellent.

"Dr. Shaw is a *Theologus* here, he has been in Africa: his *Itinerarium* is being printed; he is most excellent good company.* Dellenius, I can assure you, spoke of you, my honoured Sir, with high esteem, and promised to write immediately after my departure."

* Thomas Shaw, Theol. Prof. in Oxford, *Travels and Observations relating to several parts of Barbary and the Levant.* Oxford, 1738.

CHAPTER VIII.

INNÆUS returned to Holland with increased knowledge, and with a rich harvest of plants for the Hartcamp Gardens. His reputation was now also great in England among botanists.

More success awaited him in Holland. He was elected Corresponding Member of the French Academy of Science. Clifford and many great men vied with each other in lavishing honours and benefits upon him, so that, as he himself wrote, "he lived in the greatest affluence any mortal could wish for." He went to Leyden to hear Boerhaave whenever he pleased, and to Amsterdam as often as he chose, and thither he rode in a carriage driven by a team of four horses; and returned to Hartcamp when he liked, and there he had a man cook and many servants at his disposal, and could always receive any visitors with a splendid entertainment. But success, or enjoyment, could exercise no derogatory influence over Linnæus. He never for one moment diverged from the goal for which he found Providence had specially endowed him with his great natural gifts. His diligence was greater, if possible, than ever, keeping pace with his increasing fame. In 1736 he published *Musa Cliffortiana*, and the following year *Hortus Cliffortianus*, remarkable works which described Clifford's collections, and handed their owner's name to posterity, a testimony to the author's gratitude as well as to his rich patron's love of science. Several other great works by Linnæus were published during these years and they taxed him in many respects, as he not only carried his great reform through all branches of Natural History, but had also to invent quite a new nomenclature for the Science of Botany. Formerly many unfit and extraneous combinations had been used for the classifications and names of flowers. Linnæus could not suffer these. His simple, beautiful system demanded clear and correspondent expressions, and he coined new and significant names at the same time as he ranged the natural objects. His loving heart prompted him to use these opportunities to render homage to men, who before him had lived for science, by calling remarkable flowers by their names. But it was only after frequent importunities from Gronovius that he could be prevailed upon calling the simple little flower of the north after himself, *Linnæa Borealis*. It had formerly been known as *Campanula serpillifolia*, and the only revenge which it is known that Linnæus ever took upon his bitter adversaries in foreign lands, was to call a poisonous plant by the name of his most virulent vilifier. His incessant labour, when his great intellectual faculties were constantly in toil, and the many nights he spent poring over books and manuscripts, instead of sleeping, told sadly upon his health.

Boerhaave, Clifford and others tried every means to persuade Linnæus to remain in the service of their country. Professorial Chairs at Utrecht and Leyden were offered him, and proposals for him to undertake, at the expense of the State, scientific journeys to the Cape of Good Hope and the American Colonies. It is even related that the venerable old Boerhaave proposed his only living daughter, Johanna Maria, five years younger than Linnæus, as his bride, with the tempting dowry of a million gulden. But Linnæus waived everything; he longed to see his drooping *Linnæa* in Fahlun, and he wrote in his diary, "Everything must yield to Love."

He said that the moist climate of Holland did not agree with him, and that the Tropics, to which he had been offered to go, still less would do so, he being a son of the cold North, where the air generally is dry and crisp. But he had secret reasons of his own, for he was engaged in his own land to his "Sara Lisa Moræa."

Still, Linnæus could not immediately return home. Gronovius still desired his assistance for his work, *Flora Virginica*, and Professor van Royen solicited his aid for a new plan of the Botanical Garden in Leyden for the spring, 1738. During this time lay Boerhaave ill, without hope of recovery, suffering from dropsy. He would receive no one except his "dear Swede." When Linnæus visited him the last time, and, in taking leave, kissed the hand of the old sage, the tears pressed into Boerhaave's eyes, and his enfeebled, trembling hand brought Linnæus's hand to his lips in return, saying, "I have lived my allotted time, my dear Linnæus, and I have worked the best I have been able to do. Heaven preserve you for whom all this remains. What life has demanded of me, I have rendered, but the world expects much more from you. Farewell, farewell, my dear Linnæus."

Shortly afterwards he sent Linnæus, as a greeting of love, a copy of his splendid work on Chemistry, and not long after, the venerable Boerhaave was carried to his grave, honoured by Holland, and lamented by the whole learned world.

❀ CHAPTER IX. ❀

O return to Sweden without having visited Paris, Linnæus could not, for there were hoarded great collections by the famous botanists, Tournefort and Vaillant, men of genius, his predecessors, of whom Tournefort, like himself had discovered a new system, but which Linnæus, still more advanced and rational had superseded, but whose great labour he still highly appreciated.

There was also a world-famed library, and many scientific men of great celebrity, and an equally great attraction for Linnæus was also the famous and beautiful garden at Trianon.

He left Holland for Paris in the beginning of May, 1738, preceded by the fame of the books he had already published, which had attracted the attention of the learned world in Paris as elsewhere. He was therefore exceedingly well received and hospitably entertained by many of the great scientific men, who also afforded him the opportunity of viewing several fine herbariums and other collections, and in whose company he made excursions in the vicinity of Paris to see the growing flowers, native to the soil.

On paying a visit to the Parisian "Royal Academy of Science," he was elected a corresponding member, and the President of the Academy asked him if he would become a Frenchman, if they were to nominate him *Membrum* with a yearly pension. Linnæus refused the proffered honour and its emoluments, for love—the all-potent magnet attracted him to the north.

He was even called to the court of Louis XV. to explain to the king his sexual system, and of this occurrence has been preserved a peculiar memento in an exquisite little picture, ornamenting an elegant time-piece, to which is attached the following details which a Swedish lady, some time ago, and for many years resident in Paris, gave, relative to this royal heir-loom to Professor H. Sätherberg, of Stockholm, who has published a charming poem in twenty-five songs on the subject of the "Floral King."

"You ask me about the time-piece? With my own eyes I have seen it in the palace of the Tuilleries, both in the time of Louis Philippe, and during the Imperial reign. It stood in a little corner-room with one window looking into the courtyard and one on the quay, in the corner of the room, between the two windows. It was a small gilded timepiece in antique style. Below the face itself was a miniature picture, if I recollect aright, painted on porcelain. I cannot say if it were made at Sèvres, but it was exceedingly dainty and remarkably pretty. The little picture was not more than about six inches wide and still less in height. It represented a saloon, where in a semi-circle many ladies and gentlemen were seated in elegant costumes (court dresses of course) and in the centre stood Linnæus, also in court dress. Below the picture was an inscription, 'Linné expliquant devant la cour son systèm de —.' I don't recollect what was the word, or if it was Latin, but I fancy the meaning ended with 'fleurs.' You might understand how small the figures were, and what an exceedingly fine miniature the whole picture was; it was a *chef d'œuvre*. But it was also honoured with a place in Marie Amelie's (the queen of Louis Philippe) boudoir, and remaining in the same place in the time of the Empress. Of late years I have not seen it."

Linnæus was highly esteemed by the King of France, for when Gustavus III. was in Paris, says Linnæus' German biographer Stöver, "he was congratulated by Louis XV. for the famous man his country possessed, and he had the choicest plants gathered for him in the parks of Trianon."

Linnæus himself writes regarding this, " Carl Tr. Scheffer writes from Paris that ' during my sojourn at Versailles, the King of France has several times asked me about you, and besides the favour with which he regarded you personally, he has also, with much interest, inquired about the condition of your botanical garden. His Majesty has himself collected seeds, which he desired might reach you, and as he has expressed himself, 'I thought that this would afford the Archiater (Linnæus) pleasure.' He has commissioned me to forward them to you sir."

They consisted of 130 different kinds; the King also sent him living plants.

In another place Linnæus mentions that Louis XV. had adopted his method for the garden at Trianon. He further relates that after he had seen the most remarkable things in Paris "he went to Rouen and thence sailed with a pressing wind and storm to the Cattegat, when the wind immediately turned, and he landed at Helsingborg, whence he journeyed to visit his aged father in Stenbrohult." His mother had died in 1734, before he left for Holland. He landed on his return to his country in September, 1738, after only three years absence, during which he had become exceedingly famous abroad.

CHAPTER X.

INNÆUS drove up to the vicarage of Stenbrohult in high glee, and the meeting between father and son may easier be imagined than described. When Carl piled upon the table all the learned books he himself had written, tears of joy and pride glimmered in the old man's eyes, and again he embraced his young and famous son, who had carried with him home so many laurels peacefully won in foreign lands. With what joy his venerable father listened to the recital of his wonderfully triumphant career, how he had been received as a brother by the

most famous and learned men of the period, his patronage by the king of France, and many wondrous things he had seen in his foreign travels, all which were of the greatest contrast to the life and surroundings in the rural vicarage by Möckeln's strand. And he spoke of his hopes for the future, and of his love in distant Dalcarlia, and which made him tremble for fear the rumour which had reached him and hastened his homeward return should prove true. His stay at the present time was again therefore but brief, and he journeyed post haste to Fahlun where he found that, though the warning had been timely, yet his loved one had proved true to him, and that she with indignation had discarded the would-be suitor, Linnæus's clerical friend Brovallius, who would have played false with him, and who had now left and become professor at the University of Åbo in Finland.

The engagement of the young couple was now publicly announced, and great and universal was the joy in the house of Doctor Moræus.

However, Linnæus was compelled soon to tear himself away, and repair to Stockholm in quest of a field for his labours. He went to the capital, but soon found that his fame had not reached that city.

Linnæus writes regarding this, "Stockholm received Linnæus in September, 1738, as a perfect stranger. He intended to make a living here as a doctor, but as he was unknown to all, no one dared this year to confide his dear life into the hands of an untried doctor, nay, not even his dog, so that he often despaired of success in his own land. He, who everywhere abroad had been honoured as a *Princeps Bontanicorum*, was as a *Klimius* in his own land, as if he were come from the regions below." Had he not been in love, he would at once have gone abroad again. These reverses were a great trial for the warm-hearted man, fully conscious of his own worth, already so amply recognised in foreign countries. However, he did not quite lose heart, although to gain clients he had to frequent taverns to make acquaintances, and solicit patronage. The only encouragement Linnæus enjoyed during this unpropitious period was, that he was elected a member of the "Society of Science" at Upsala. Gradually, however, a change for the better took place, and his callous countrymen at last awoke to a recognition of his great merits, and the number of his patients increased daily, until they soon became very numerous.

Swen Hedin wrote in his Academical panegyric, " Linnæus was forced for a long time to renounce his devotion to flowers and instead serve under the ensign of Esculapius. He now needs must turn his eyes from what was beautiful, and full of life in nature, to succour suffering and agonised humanity. However, he fulfilled his duties with great success, and he was sought even by the highest functionaries of the Court. Linnæus gradually became acquainted with many learned and eminent men. The Royal Academy of Science in Stockholm was founded in 1739 by Captain Triewald, in conjunction with the members of the nobility, Höpken, Bjelke, Cederhjelm, and the zealous Swedish patriot, Jonas Ahlström, (creator of a native industry of woollen manufactures), and of this Academy Linnæus was made first President. Shortly afterwards he was sent for by Count Carl Gustaf Tessin, the Speaker in the House of Nobles, one of the leading spirits during the short time Sweden was a Republic, and a great patron of Science and Art. The noble Count inquired of Linnæus if he had no supplication to make at the Diet. 'I feel convinced,' said Tessin, 'that the Diet will think it a pleasure to favour a Swede who has distinguished himself so much abroad.' Linnæus answered that he did not know of any solicitation to make.

"'Consider it well, however, Doctor Linnæus,' replied the Count, 'think the matter over until tomorrow, and then come back to me again.' Linnæus met Triewald in the afternoon, and he advised him to seek to become appointed lecturer in Mineralogy in Stockholm, now vacant, with the remuneration of 100 ducats, and which place had formerly been filled by Triewald. The following morning Linnæus returned to the Count, who with a complacent smile received his petition, and asked him to return to dinner. Count Tessin then went to attend a meeting at what was called the 'Secret Committee,' and when he returned he met Linnæus with eyes beaming with delight. 'I congratulate you, doctor,' he said, 'your petition has been granted by the Diet.'"

For the 100 ducats it became Linnæus' duty to lecture during the winter about the mineral and geological collections in the capital; during the summer months to lecture in Botany, the *locale* of which was assigned to no less an honoured place than the House of Nobles.

In 1739, Linnæus, through the intercession of Count Tessin, was also appointed physician to the "Royal Admiralty."

The great Count Tessin became his patron, and gave Linnæus for a free residence, those of his own apartments which he had lived in before his marriage, and Linnæus also daily dined with the Count, at whose table he met many of the great and influential men of the Republican period. His practice was now so great, that he alone had more to do than all the other physicians of the capital together, and his yearly income rose to "9,000 daler copper-coin." He then thought the time had come for him to complete his good fortune by marrying his beloved and faithful Sara Lisa Moræa, and that union was celebrated the 26th June, 1739, at the country seat of his father-in-law, which was called Sweden, and situated near Fahlun. A month afterwards he brought his young wife to their own home in Stockholm. Linnæus received a letter from the famous Professor Albrecht von Haller, in Göttingen, dated the 24th November, 1738. Of course the letter was written in Latin, in which all Linnæus' foreign correspondence was carried on. It said, "you, from whom Flora expects more than of any other botanist, I pray to take advantage of your fortunate position, and return to a milder climate. In case my own country calls me back, and I hope it will soon happen, I have decided that you, provided it meets with your approval, should inherit the botanical garden here, and other honours, and I have already spoken with those who have to decide respecting all these." And he again wrote the 19th January, 1739, "My decision regarding relinquishing the botanical garden, is the same as before. I remain here only a few years, and can leave it to no worthier than you."

Linnæus' reply dated the 12th September, 1739,

expresses the greatest gratitude for the offer, "I may say," he writes, "that I have numerous acquaintances amongst my fellow beings, and many have been attached to me by friendship, but no one has ever made me such a handsome offer as you have. I can give you no other answer, seeing that you have put yourself in the place of a father to me, than to render you a short account of my life till this hour." He then briefly recounted his career, finishing by relating that he had recently been married. Thus settled in Sweden he could not think of a professorial chair in a foreign land, but he proposed to come on some future time on a visit to von Haller, and bring his "darling wife" with him. "To reside abroad," he says in his Diary, "from that time never more entered my mind."

Von Haller in his "Bibliotheca Botanica," acknowledges the Linnæan system as a new era in Botany, and without prejudice reckons his own works as pertaining to a past period. What von Haller thought of Linnæus may briefly be gathered from the following extract from his Tome X. "Linnæus and his Contemporaries."

"In the year 1732, Carl Linnæus' first treatise appeared, (Florula Lapponica) of a man, who produced the greatest change in the universal herbarium, and who at his death possessed it almost in its entirety. By nature endowed with an ardent mind, quick imagination and systematic genius, he made use of abundant opportunities, especially in the latter part of his life, when on all sides from everybody, treasures of nature flowed in to him; he laid himself out in the new herbærian plan with all the strength of his mind, which he possessed in a great degree, while he was living and witness to it. His things were pleasing to and received by many of his contemporaries, nor can it be denied that the different parts of plants are defined by him with much more accuracy than was formerly usual, and it must be acknowledged that he expressed nature much better, of which descriptions are now given, even as if an almost new language had been made for this subject."

CHAPTER IX.

"Gracious Sirs! Gentlemen!

When I compare those sciences which flourished 100 years ago, and now soon have disappeared under their horizon, with these which now rise with the dawn of time, I find as great a difference between them, as between night and day. When I read the academical works which a century ago were published, I find in them an incredible deal of reading, with more labour than in ours. But when I have read a few sheets and reflect on what I have read, I notice a whole mass of chaff with but few good ears of corn. Everything considered in the dead languages, in the erudition of the ancients, or *placitis*.

"When, on the contrary, I turn myself to the approaching time, I see only the practical and fruitful sciences, and that which forms their foundation. I see sheer Natural History, Astronomy, Physics, Mechanics, Social Economy, Chemistry, Medicine, Arts, Manufactures, &c., which reflect through man the wisdom and omnipotence of the Creator, as we are being solaced and sustained by this work. It must have proceeded with sciences like with plants. The plants begin from the smallest germs, grow to leaves and stalks, but at last bloom and bear fruit.

"He who would read sciences, must first learn the alphabet, then spelling, then learn phrases and glossary; at last he understands the text. But to come to my own subject or natural science, I find the same growth in it. The ancients confined themselves to the enumeration and specification of stones, plants, animals; more they did not accomplish; it was all a higgledy-piggledy work, without their perceiving whence the road was to lead; now after having trudged so long, we begin to behold the palace of nature, and to look into its formerly closed chambers. Then we discern, in the greatest confusion of the created things, the greatest order of the Creator, nature's economy, nature's policy; the Creator's miracles, in fine all, even the most contemptible, and the most poisonous are employed for man's pleasure, use and necessities where everything is ordained to some certain purpose, and where nothing becomes unnecessary of all which God has granted us in the three realms of nature. But that I may farther follow the progress of my sciences, I find the earliest according to *palingenesia literarum*, worked to discover the created things, and to carve them into figures and descriptions. The descriptions were then wrapped in long and diffuse orations which now are left as naked as they were born, with as many words as are significant, without twaddle, that is with characteristic descriptions, which exclude the common natural structure; it is now no longer said about animals, that the head is placed before the body, or the eyes in the head, or that a bird has two wings, and two feet; neither regarding plants that the root is dark, the leaves green, the flower pretty, or that the fruit succeeds the blooming, but only bring forth traits of distinction.

"With the figures in the same manner. If I compare the first figures *ex gr.* *Cubæ*, in the *Horto sanitatis* with *Ehret's* in *horto Cliffortiana*, the former might be regarded as goblins compared with angels.

"The figures in the fifteenth *seculo* required superscription, if they were to be recognised. A hare was painted and above it was written that it was to represent a bear. They made pictures after other people's description, without seeing the object, so that one now only feels loathing for the book from seeing the figures.

"In the beginning of the sixteenth *seculo* the figures were still wretched, but before the end of it, they became more generally endurable, particularly since good masters began cutting in box-wood.

"At the commencement of the seventeenth *seculo*, when they saw that the fine lines could not so easily be incised in wood, some people began pointing them in copper; it went somewhat slowly in the beginning, and the pictures became dirty; but before the end of the *seculo* appeared Dodart's splendid figures in the documents of the French Academy of Sciences, not to speak of others.

"Still at the beginning of this *seculo*, although the figures became passable, it was thought necessary to accumulate a great number of synonyms every time any plant or animal was to be defined, so that every one was compelled to acquire a great library, and refer to all quotations, until the Natural Historians began painting their objects with lively colours, so that no mistakes possibly could be made. I do not here speak of the illuminated figures of the ancients, who with one colour made all leaves green, or all yellow flowers equally yellow, but I point to the insects from Surinam by *Meriana*, natural objects of *Seba*, birds of *Frichen*, fishes of *Catesby*, plants of *Ehret*, insects of *Roesel*, birds of *Edwards*, shells of *Regenfus*, where the object appears as living, as the best portrait-painter can render a human face. Amongst all these excels *Roesel* in insects, *Edwards* in birds, *Regenfus* in conches, *Ehret* in plants, which are so splendid that the most callous Hottentot might be induced to love and admire the works of the Creator. When I farther inquire what cause has driven this class of literature to such a height, I find that encouragement alone has done it all.

"Rich English gentlemen contributed to *Catesby's* journey to America, and paid his tables with high prices, *Roesel* was encouraged by *Baron* to his work, *Ehret's* tables were paid with a guinea a piece as fast as he could produce them. His Majesty of Denmark liberally got us *Regenfus's* shells. How *Edwards* was encouraged may be read in his preface. Now boast the English of their *Edwards*, the Germans their *Roesel*, the Danes of their *Regenfus*, and that with good reason. Our libraries have, through them, become more illustrious, the potentates themselves must tarry when these books are being opened, and are induced to patronize the sciences. But, gentlemen, let us compare the miracles of our time with the figures which your *Clerck* produced of the choicest Indian Papilions, and you will see gentlemen, that blackest envy must admit that none of the above mentioned, yea, none in the whole world has been able to present any book with figures so rare, gorgeous, vivid, and beautiful, as can equal *Herr Clerck's*. This has all the more astounded me because, for twenty-seven years, since I returned from my foreign travels I never have been able to procure a single proper figure. If I had had this aid of a man who would have had the opportunity of seeing more than anyone else of rare animals and plants, I would have been sure to hatch, every quarter, some observations. It almost aggravates me that a man in such an humble sphere, without the slightest granting, with nothing but an ardent desire to do honour to his country, has been able to go beyond all others; but when I consider his slender means, it cuts me to the heart to see him, who has consecrated himself to sciences, being blighted by negligence, without encouragement, without gratitude.

"I understand, gentlemen, that you know his situation better than I, for he is your associated member. If my intercession will create any commiseration for him, I offer it from the depth of my heart.

"I feel certain that you, gentlemen, who constitute such an illustrious Academy of Science will not content yourselves with only receiving, furthering, and publishing results, but far more encourage and promote science. What more worthy subject could I present to you? How would not our country be honoured if his book became general? How indigent would it not appear that such a work was produced, but could not become common, because of the callousness for science in our cold north? But to furnish proposals for relief, *sic stantibus rebus*, is somewhat more difficult.

"The Academy of Science consists of two kinds of members, of workers and magnates. To the former we ascribe all the results which shine in the documents of the Academy, it is the duty of the latter to encourage and promote. The magnates might become as useful, yes, even more so, than the most diligent workers, if only the Academy would afford them opportunity. The majority of the magnates would, with delight, contribute to this with the fulness of heart; I have seen and found the most evident proofs of this in respect to myself. These magnates are the main springs of the whole realm; what would it cost a whole kingdom to appoint the assiduous *Clerck* to some advantageous situation? It depends upon you, gentlemen, to intercede with a few words to the magnates, and I feel certain that many a one amongst them with delight would become a Mæcenas.

dixi

" Upsala, 1765, the 3rd January,

"CARL v. LINNÉ."

To the Royal Academy of Science, Stockholm.

" When I was in Paris I travelled with Professor Jussien into Burgundy, far beyond Fontainebleau, to see the extraordinary orchids, *Ex. gr., orchis mirscam referens, orchis hiante arcullo, orchis fl. conglomerats, albo, punctato, etc.*

" The very first day I came to Öland, (a Swedish Island in the Baltic,) I found all these grew wild in the meadows, which I had written for to Spain, and obtained last winter through the French Minister. If the whole world had told me that they grew on Öland I would never have believed it, now I believe it. The 'Tok' of Öland is *Pentaphylloides, Chorancense,* which in the whole world has never before existed anywhere but in one place in Scotland and in Siberia, whence I got it of Professor Amman in St. Petersburg; here it grows everywhere.

"Öl., Runsten, 1741, June 9th,
"C. L."

CHAPTER XII.

THE Chair of Botany became vacant in 1740, through the demise of Olof Rudbeck. Linnæus was proposed to fill it, but again intrigue was busy to exclude him, and his secret enemies succeeded in getting his old antagonist, Rosén, promoted to the professorship. Soon afterwards Professor Roberg resigned, and there was none more eligible to fill the vacancy than Linnæus, but the opposing faction, who wanted to exclude him from Upsala, had nearly succeeded in their object, if the delegates of the Diet themselves had not taken the matter into their own hands. The Consistory was peremptorily commanded to propose Linnæus for the professorship, and thus at last, in 1741, he was appointed, but got the Chair of Medicine and Anatomy, which was not the one he desired.

Before he entered upon his new duties, he was requested by the Diet to make a voyage to the two islands of Gothland and Öland, Swedish possessions in the Baltic, to examine the natural history objects of these places.

On his return, in the autumn, he removed to Upsala, where he was watched by jealous eyes, but where he was to find his proper field in life. Rosén and Linnæus agreed to exchange their respective professorial chairs, that each might become filled by the right man, for Rosén was acknowledged to be great in Anatomy, while he could in no way compare with Linnæus in Botany. This exchange took place in 1742, and from that moment Linnæus became his country's pride. To study Botany in Upsala, under Linnæus, became an ambition to many. Disciples came from many lands, some of whom paid large sums for the privilege of listening a few hours to his lectures. Seldom had any foreigners come to this University of the north before, hut the fame of Linnæus attracted many, amongst whom Dr. Kane, all the way from America; the two Princes Demidoff from Russia, and Lord Baltimore from England. The latter sent Linnæus for a lecture, one forenoon, a splendid little box of gold, containing 100 ducats, and also a fine *necessaire* of gilt silver, weighing 6-lbs., and beautifully wrought. But then, never before had science been expounded with such charms from the lips of any academical lecturer; for Linnæus held his audience perfectly spell-bound by his lucid and eloquent exposition of the subject in hand. The sacred fire which warmed the High Priest of Nature, also kindled a kindred flame in the hearts of his hearers, until a general love for Botany and Natural History prevailed everywhere amongst the numerous students of the Upsala University. His lectures also bore on Medicine and Sanitary Science.

Several of his disciples were sent on journeys of scientific researches to many distant lands, where they had opportunities of zealously devoting themselves to gathering rare collections, and bringing into notice many curious and useful things, thus gaining honour for themselves, their preceptor, and their country.

An unusually strong winter and various other, untoward circumstances, had combined in playing sad havoc with the Botanical Garden, and on Linnæus entering into office it became his first care—in conjunction with the the Chancellor of the University, Count Gyllenborg, and Baron Hårleman—to restore the plantation. More than once Linnæus thought of how the wish of his childhood, to become gardener of this famous place, had been more than realized: now, that through his widespread connection with all the greatest Botanists and Horticulturists in the world, he could in a short time transform the despoiled Garden of the University to a Paradise like that one at Hartcamp. Its area was greatly enlarged; well appointed conservatories were erected; and everything that art, taste, and great means could procure was affectionately lavished on the Botanical Garden.

The Floral King.

The residence allotted to Linnæus, as Professor of Botany, and which formerly had been inhabited by Olof Rudbeck, adjoined the plantation. "It was really an old owl's nest," says Linnæus, but he soon transformed it into a charming and comfortable home, where, with his beloved Sara Lisa, he passed many years of happiness, and where they saw their children grow up, young saplings in their domestic garden.

In 1749 Linnæus was appointed Archiater, and in 1762 he was ennobled, when for his coat of arms he chose three crowns on three different fields, emblematic of the three Kingdoms of Nature which he had systematically arranged. The helmet was adorned with his favourite flower *Linea Borealis*, and from that time he wrote himself von Linné. He was also decorated with the "Star of the North," in those days a great distinction, and several medals by the Academy of Science, and other learned Societies, were struck in his honour. He was made a member of no less than eighteen Scientific Societies in the world.

Ferdinand VI. of Spain invited Linnæus to settle

in his realm, proffering a patent of nobility, 2,000 piastres as a salary, and free maintenance, and liberty to retain and exercise his Lutheran persuasion in Catholic Spain.

Also Catharine II. of Russia made him a tempting and brilliant offer to reside in her realm, and the Royalty of his own country embraced every opportunity of showing their delight with their great and famous subject.

. . . .

All leisure time which Linnæus' friend, Dr. Hagström, could possibly snatch from his official duties he devoted to the study of bees, and how pleased Linnæus was with his work his own words testify. He wrote, 28th October, 1768, to Dr. Hagström, relative to his *Pan apum*, or the food of bees, which the Royal Academy of Science caused to be published: "I have eight times read your *Pan apum*, which you kindly sent me; I admit, without hypocrisy, that it is a gem. Your perfectly novel results, brief and ingenious style, clear thoughts and inferences, make me think that he must have a heart of stone who would not be affected by, and conceive a love for the book, however unlettered the reader might be. You, Doctor, have with this sole thing, graven your name in the endurance of time, which no changes of the future can erase. I congratulate you, Doctor, to immortality.

"I remain, yours,
"C. v. LINNÉ."

. . . .

"Someone told me yesterday that the devil would take me, because in my journey through Skåne I had not mentioned all those with their knightly titles and honours I had referred to of that ilk! I did not know, and still do not know, if it is necessary, at least for me, who am sparse of all titles, except my own; be so kind as to ask my fatherly friend regarding this, what is his opinion, for I know he does not want the devil to take me. Is it thought quite essential? Then the thing may be amended, if I make a catalogue of the Knights mentioned in this work. But it is never the custom abroad.

"*Honores populi nostri quondam fuese rariores et ob eandem caussam gloriosi, &c.*

"Favour me with a reply to this, that this matter for my conscience may not gnaw at my heart.

"Here is a theatre by Nature; miracles which might occupy the greatest *Physicus* for long time, and also please a *Lithologus* who had a mind to collect various petrifications and rare *Ostracodermata*, which I was compelled to leave for those who had time to remain longer in this locality. When we now further consider how so many strange animals have come to be buried in this Bal's mountain, and which scarcely now are to be met with in Europe, we encounter a new argument which claims no less consideration.

"*Tertacea*, or the whole shell species, which is located at the bottom of the sea, is divided into *Littoralia* and *Pelagica*. Shell collectors call those mussels and shell-creatures *littorales*, which do not keep to the deep, but only close to the land, so that their shells are thrown upon the shores as soon as they die and perish, from which reason these shells are common in natural collections.

"*Pelagica*, on the contrary, are those shell-creatures which keep in the depth of the sea, and never come near the shores; therefore their shells are very difficult to obtain.

"The depth of the sea is mostly sterile, and covered with sand or corals, without much fish, creatures or plants, for where no plants exist there are seldom any worms or fishes, which both live by such. We have thus not more than one single plant which can grow in the greatest depths of the sea, which is called *Sargazo*, and there is no other herb so plentiful in the world. This floats on the water and affixes itself together, so that the sea, at a long distance off, looks like a green meadow; under this keep the rarest creatures and shells, or *Testacea plelaica*, which, as they gradually increase, also gradually die away, when their shells fall to the bottom and fill it. The most of the *Testacea* in our mountain referred to, are *Petagicia*, and must have increased there where a *Sargazo* has grown; but how they have come here in this country, is a more difficult knot to unravel. The majority assert that the shells have been brought here at the Great Flood, and that they thus bear witness to the changes of this wondrous earth. But those who insist on this seem to me to be very little at home in Mathematics, for how could a weltering flood possibly throw the shells some thousands of miles away to a certain place, and then place the other alluvial layers in such a perfect order? If we consider these phenomena carefully, one must perforce admit that the earth has lain under water and sea, and that in this locality then *Sargazo* existed, below which these creatures have lived and died; on which, at last, when the water had diminished, and *Sargazo* driven away, gravel has been thrown up by the billows on to the new bank and strand, and which has grown together to stone. Thus we see here the rarest shells in great abundance; the plainest evidence of landings; how many infinite thousands of life Nature here must have produced before she could fill this small locality. C. L." *Journey through Skåne.* (South of Sweden.)

"From this can also reasonably be inferred that many new lands and mountains formerly must have originated, to which *Sargazo* and *Fucus* seem to have contributed the most. This is a plant existing on the trackless ocean, and it might almost be said the most abundant of all plants in the world. It floats on the water, and is best known to them who have visited the Indies, where it frequently covers the sea for hundreds of miles, that it looks like a green meadow. We know that where the aquatic plants float with their leaves on the seas, reigns a constant calm, just as it takes place when the whalers throw oil on the boisterous billows, which then immediately subside. The same way where *Sargazo* floats is a continuous calm, so that *mare pacificum* from that got its name.

"When water is allowed to remain still, it deposits its sediment at the bottom, which forms clay, consequently, where *Sargazo* lies in constant repose, the water deposits the clay in great abundance, and fills the bottom. Among the *Sargazo* exists other kinds of birds, other kinds of fishes, viz.: those that have floating eggs, and other kinds, *Vermes*, *Cockles*, *Conches*, *Medusæ*, &c., than those which are known near the shores. These gradually die, when their bodies sink down and are mixed with the clay. When such are mixed with the clay which themselves are covered with chalk shells, they transform it to chalkstone. By this can be explained how in the chalk hills on the wolds of Öland have been found such petrification, *Orthocreotes* and others, the animals of which now are perfectly unknown. To this seems also *Kinekulle* give occasion, which has likely been originated in this manner, that the lowermost layer of sandstone has been amalgamated by sea sand, on the top of which slate exists, generated by the black mould which covers the sea sand at the bottom. On the top of this is a thick layer of chalkstone, full of unknown petrifications, may-hap let down *Sargazo*. Again on the top of this lies slate of decayed *Sargazo*, the top part is grey stones, which are generated from gravel, and probably is the same which was thrown up by the sea, when the mountains began to form a shore.

" It is remarkable that the uppermost layers in all mountains only consist of greystone. From this is inferred that they are not come thither from the beginning, because the layer which lies next below is slate-stone, which always is generated by black mould, and as all mould is formed of vegetables, there must have been herbs before the greystone layer had been added, which consequently cannot have been created. Those who will ascribe all this to the great flood, think but little, and see still less.

"A much longer time has been required to this than that lasted.

"*Oconomea Naturæ*, or the Creator's all-wise arrangements on our earth, observed in the contemplation of the Three Kingdoms of Nature, regarding their propagation, maintenance and decay."

✤ CHAPTER XIII. ✤

"A STUDENT, Dan Rolander who, a couple of years ago, learned a little from me about insects, communicates to me two observations, which perfectly astounded me, and I requested him to send them to the Academy of Science, because they were most marvellous, and such as I had not expected from a Réaumar or a De Geer.

"One of the insects possesses the faculty of shooting with his back, and of popping it off with a bluish smoke, so that all the other insects that chase it get frightened—and this quite twenty times in succession when ever it please. This is a phenomenon, which no mortal has ever seen or heard of, so that if it were from India, it would stagger us, and yet, though it is common in our forests, no European knows of it.

"The other insect has on its forelegs two bowls, which are folded together under its chin, like a hemisphere, I have often seen it, but have never understood it. In these bowls it gathers like bees, *pulverem antherarum*, but the bowls are perforated with innumerable holes, so tiny that they can hardly be seen by the naked eye, the insect sifts the flower through these and takes thus the broken pollen, leaving that which still serves the plants for the fructification; incomparably curious.

"Since we have first obtaind *inventa* about these two insects, it remains for all the curious to make a hundred trials in this respect, and new discoveries, but it is yet his *gloria* who first detected it.

"When this comes to the Academy of Science, I should advise that an *inventor* be distinguished from *compilers*, for where we have one inventor, we have 1,000 compilers."

In the beginning of the year 1761, Linnæus wrote to his friend, P. W. Wargentin, the Secretary of the Royal Academy of Science :—

"DEAR SIR,

"I take the liberty of communicating to you a phenomenon, which does not belong to my *forum*; I hope, however, that you will excuse me, sir, if I go *extra oleas*.

"The ancients have spoken about an art, which now is *inter artes depreditas*, which they called *palingensie*. *Francus* has searched through the erudition of the ancients for all the *dicta* which exist regarding it, to prove that such an art was known to them. Since *palingenesiam literarum* I know no one who has understood it. It has been said that *Kirkems* showed it in Rome to Queen Christina and her suite. I read, about twenty years ago, in *Actis Nat. Curios*, about someone who had prepared an *infusum fugidum* of roses, which had not succeeded, he put the glass on a shelf where it remained forgotten for some years. One evening he got up to the shelf, he saw in the glass that the humours had parted themselves into something like rose-leaves, but without colour; he made a drawing of them.

"I received a month ago, from the Councillor of Commerce, the Honorary Herr Bungencrona, a quantity of tea-plant seeds. I tried them in water to see if they were sound, but found that they were decayed, although the kernel appeared sound, which generally happens with the seeds of the tea-plant. I poured water from the water-jug into the hand-basin in which the seeds lay, and macerated for eight entire days, the water became brown, the seeds were taken away and sown. This brown water remained another eight days, if not more. I found great pleasure in observing how the brown water separated itself from the clear water in the hand-basin, and looked like a painting of brown shrubs in the liquid water, and I thought I saw here a species of *palingenesie*. At last the water froze in the cold room, and perfectly retained the figure which the tinged water had before, so that the ice lay in the hand-basin like branches and leaves. The ice was about an inch thick, and between the branches the water had not formed the slightest ice. It is very strange that I have never seen anything similar. I showed it to Herr Adjunctus Melander and Magister Docens Bergman, who both viewed it with the same astonishment. The ice figures, which show themselves on the windows, are flat and filled up between the branches with ice. There have been those who have thought that this comes from vegetable exhalations, perhaps, after they have passed through the bodies of animals. It is noteworthy that the water which was in the water-jug was also frozen, but as no tea-plant seeds had been soaked in it, it had frozen in the regular way, according to the laws of crystalization *ad angulos*, as salts are crystalized, for which reason Newton says that water is a liquid salt.

"Please excuse my troubling you with a thing which I do not know myself whether it is important or not, for this is not my business.

"C. LINNÆUS."

EXTRACT OF A LETTER FROM PROFESSOR LINNÆUS, TO THE SECRETARY, HERR ELOIUS.

"You will call to mind, sir, how often observations have been sent in from the country to the Royal

The Floral King. 143

Academy of Science, that *water has been changed into blood*, and how frequently I have pronounced my opinion against this. It is difficult to say anything in this matter since it is known how the defunct Bishop Svedberg espouses this when he calls it an *Abyssum Satanæ*; he says, positively, that *it is not anything natural, and that when God allows such a miracle to take place, Satan endeavours, his tools also, the ungodly, self-relying, self-sufficient, and wordly people to make it signify nothing.* Regarding this see also Archiat Hiener's (Urban Hjärne) Flock on Water, who has various tales about water being turned into blood. It is not only with us that water has been turned into blood, it has happened in France (Svamm quar Fo) in Holland and Leyden, that the people were appalled at the water being turned into blood. In England, Durham, etc., and Sweden it is more common than anywhere else. Here in the University garden are three ponds, the largest of which is free from aquatic plants, every summer at the solstice turns into blood, and that every morning and evening when it is calm. This water-blood is strange in two ways; I have shown it, amongst others, to our erudite physicus, professor Klingenstyerna, who viewed it with pleasure. Every morning when it is calm, something resembling black powder lies on it, strewn evenly along all the edges, this black powder moves itself gradually from the edge closer to the *centrum*, as if it were an army under command, so orderly it proceeds, until after a few hours it stops and is gathered together *in centro*. On the water where this powder has passed, seems a grey, almost invisible, film to flicker, I cannot say by what produced. If anyone takes up some of this powder with a spoon it will be seen how it jumps, it is all alive and consists of many millions of little insects, which Herr De Geer has so incomparably described and drawn under the name of *podura aquatica*. At the same time something resembling clots of blood is seen down in the water itself, in the same manner as when anyone allows a vein of his foot to be cut, and afterwards puts it in a vessel with water; these blood-clots make the water quite red, or the colour of life in the places where they are; they are sometimes thicker and denser, dissolve and quite disappear, when others in lieu rise. All the water in the pond is so full of these that no one could use it for cooking; towards nine or ten o'clock in the day it vanishes and dissolves, but towards evening it returns, and even early in the morning it makes its appearance, particularly if rain has fallen during the night. If one takes any of this with a spoon it is found to consist of so many millions of little creatures, which each resembles a small grain, the size of a tiny midget, with two long forked horns, by means of which it throws itself up, and it has one eye in its forehead; it is called in Latin *monoculus*, and is well drawn by Swammeri, quar, p. 66 + 1.

"When water is stagnant, it becomes rotten, and muddy, by which means food is generated for these creatures; when they get sufficient food they increase incredibly, just as vermin in children's heads when they have sores. It is certainly a miracle that so many millions can so suddenly be propagated, which shows indubitably the all-wise power of The Infinite, but that it should mean much harm for the country, I cannot see that it follows, more than if anyone chose to say that because there are many fleas in an unclean cattle shed, there will be no sledging in the winter.

"Nor have we any example that these little creatures do any harm; tame and wild ducks, the water shrobba, (Dytiscus), water-bugs, (Cimex Tipula), water vantsor, (Notoneitæ), all make their food of them.

"When one goes a long sea voyage, it often happens that the drinking water on board becomes full of them, but when either a few drops of wine or brandy are poured into a mug of the water, they immediately die, and sink to the bottom."

Bishop Svedberg, referred to in this letter of Linnæus, was born in 1653, on the small country estate of *Sweden*, close to Falun in Dalcarlia, which afterwards came into the possession of Assessor Moræus, the father-in-law of Linnæus. Jesper Svedberg was a man of great endowments, and deserves a place in the annals of his country; his life was in every respect a worthy example to follow. His ardent faith, however, could not wholly free itself from the superstitions of his time, and he often related in his sermons, about persons having seen apparitions, and spoke of prophecies by then living persons, with that warmth in the recital which only a sincere belief in them gives. In fact in thoughts spiritual and supernatural, he was the precursor of his celebrated son, Emmanuel Swedenborg, whose creed and teachings about the secrets of the spirit-world, and the New Heavenly Jerusalem, founded the Swedenborgian sect in England, but which has only recently been formed into a sect in Sweden. This, however, did not prevent Bishop Svedberg from joining, with zeal and energy, in the work by which, at that time, the Swedish church was regenerated. His celebrated son, Emmanuel Swedenborg, lies buried in the Swedish Church in London.

"The only thing which has caused *historia naturalis* to be applied to economy has been that it has been applied according to the economy of nature, which consists of three realms; and as they never have been treated by one man, but have always been separated, there has never been any link. Now since mineralogy, and soils, etc., are brought to different professions, there will be different *principia*, and no link, so that what was thought would have aided the cause, is just that which ruins it, and that as truly as I hope to be saved, remember my words in the future."

UPSALA, 22ND MARCH, 1751, TO WARGENTIN, THE SECRETARY OF THE ROYAL ACADEMY OF SCIENCE.

. . . .

"Baron Müchhausen wrote ten years ago that the seeds of mushrooms are alive; I do not know anyone who has made an observation about this. Just now I have seeds of mildewed ears and mushrooms in water, and I have watched them gambol, like fishes in the spawning season—before they fix themselves and are transformed into mushrooms. A strange and wondrous metamorphosis! *Quis nisi vidisset crederet?* I have shown them alive to all my disciples who have been with me. I verify *Plinus: mihi contuenti sese persuasit rerum natura nihil incredible existimare de ea.*

"In insects we often see caterpillars so dormant that they can scarcely move, yet they transform into the most lively creatures, and here we see most lively creatures transformed into motionless and almost lifeless mushrooms. Had Leuwenhock seen these he would have gained strong argument for his *vermiculis spermaticis* and thought that human beings grew like mushrooms. Now from these, *vermicuti infusorii* are supposed to generate; we shall see if in the end, all fermentation turns into sheer living particles, *i.e.*, *medullares substantiæ absque corticali*.

"A couple of months ago, I got from the Mediterranean, 300 rare insects, I have written to all the people I know out there, but cannot get to know who has sent them."

UPSALA, THE 18TH NOVEMBER, 1746, TO P. ELVIUS, SECRETARY OF THE ROYAL ACADEMY OF SCIENCE.

"I confess that I find it very difficult to give any opinion about the bread baked with reindeer moss; it certainly would be an incomparable thing in times of dearth, I am sure that Assessor Hesselius means as well as anyone, but it goes too much against my ideas.

"1. Because in the entire *familiæ* and arts of mosses exists no *esculentum* for people, and scarcely any for quadrupeds, for regarding heath-moss, it must be boiled with milk, or else it will purge, which proves that it contains something not very welcome for nature.

"2. When I tasted the flour for the bread at Assesssor Hesselius (when I was in Örebro) I noticed that some moss-flour had a nauseous taste which long stayed on my tongue, and which was by no means a good sign of any *aliment*, all the more as the Creator has put taste and smell as chief physicians for all men and animals. One must verily proceed *cant*, when it concerns so many people's lives. One ought neither to throw to the winds a means, which could sustain the lives of so many people; with the like it is best not to be in haste. One ought to rely in this, as in everything else, upon experience. It would be worth while to gather this in large quantities, and have much bread baked from it, ask dogs, pigs and other animals, who have pure taste, how they like it; allow poor people to get it gratis, yet proceed slowly, so that one is enabled to check it where it might harm.

"The reindeer, who has got this of the Creator for its principal food, possesses no gall-bladder, and fain passes it in the summer, when it has got something else.

"The cows can eat it during winter, after it has been moistened with warm water, but still they would rather eat straw, and the milk becomes insipid (*sapor fatuus*) from it, which any peasant in *Vesterbotten* knows."

Linnæus put great faith in the taste and smell of medicinal remedies, and two treatises were published under his præsidium, probably written by himself *Sapor Medicamentorum* (1751) and *Odores Medicamentorum* (1752).

TO P. ELVIUS, UPSALA, 1747.

"DEAR SIR,

"I had a letter by to-day's mail from *Med. Studiosus*, Hasselqvist, who entertains some hope that he may be sent to *Terram Sanctam*. I would fain write for him to our *Mæcenates Scientiarum*, to the Academy, to you, and others, but truly I feel ashamed, who have never learnt to be impudent; I feel also myself an abhorence for those who are never content. I have got enough these times from *publicum*, when I got Kalm to Canada; God grant that journey may turn out well, we shall then afterwards get more assistance from *publicum* than we would require.

"But to return to Hasselqvist, he is the most efficient, and of whom our faculty can entertain the greatest expectation. I know no one, who more earnestly, more continuously, and more diligently worked with all that is curious than he has done, added to this he is rather poor, but humble, honest, smart, and desires to go.

"You know, sir, how much the greatest *Patres Theologi* have made their best endeavours for a correct version of the Holy Writ, and that they have had the greatest trouble with *animalilcus* and *plantes Biblicis*, how impossible it has been for them to make this clear, as long as they did not know what there existed, living and growing in the Holy Land.

"Soon all the realms in the world have been explored by *Botanicis*, but to this day never has any *Botanicus* been to *Terra Sancta*, for neither *Bellonius* nor *Rauwolfius*, nor *Shaw* did know any plants, which can be seen by their *itinerariis*. I think, however, if anyone went thither, who there defined all that lived and grew, how easy it would afterwards be in *philologiea sacra*, what great name would he not get, who by the *theologis* of all nations were extolled with equal praise; I doubt any erudite man could be exalted so high. Verily, had not my years, appointments, and wife made me so delicate, I had, myself, great desire to go thither.

"I admit that such a journey would add nothing to the well-being of the realm, but still it could be becoming for the nation, all the more as the young man does not ask more than a *regium stipendium* from each faculty God grant that eight scholarships yea, eighty could render as much lustre as these four. He risks his life, *publicum* scarcely anything. He will procure something from the clergy of his diocese, I shall also give him as much as I can manage.

"Herr Kjerman and Plomgren could help him on his journey to Smyrna, thence to Joppa is not a long journey, thence to *Terra Sancta* not far. The young man must go as Pilgrim, without any show.

The whole journey could be effected with little, but not paid with much. If you, sir, can assist him, or somewhat recommend him, do so, for he is worthy of your assistance, and of that of us all; he asks for no capital.

"I see with wonder how our nation endeavours every way to become illustrious in the world. I dare not ask a single person more in this respect, so that I may not prove tiresome.

"*Vale,*

"Upsala, 1747,

"CARL LINNÆUS."

Fredrik Hasselqvist, born 1722, student, 1741, disputed *pro exerc.* under Linnæus præs, 1747, "*de viribus plantarum,*" sailed August, 1749, from Stockholm direct to Smyrna, where his relative, And. Rydelius, was Swedish Consul-General. After an excursion in the Spring of 1750, in Natolia to *Magnesia ad Sipylum,* he proceeded in May through Alexandria and Loretto to Cairo, where he remained almost a whole year. Since more money had been collected by voluntary contributions in Sweden, for the continuation of his travels, he, in March 1751, again proceeded from Damiette to Jaffa, Palestine and Phenicia: then from Sidon, crossing by Cyprus, Rhodes and Chios, returning to Smyrna, where he arrived in August of the same year with a great collection of all kinds of objects of Natural History, but also a prey to deep consumption, to which he succumbed in the village of Bagda, near Smyrna, the 9th of February, 1752.

His valuable collections were seized for debt, but through the instrumentality of Linnæus, were redeemed with 14,000 daler copper coin, paid by Queen Lovisa Ulrica, the then reigning sovereign of Sweden. Arrived in Sweden they were added to the collections of Natural History at Drottningholm, a royal residence on an island in the lake of Mälar, some six English miles distant from Stockholm, and called the "Versailles" of Sweden. They were described in Fredrik Hasselqvist's *Iter Palestinum*, or Journey to the Holy Land, by Her Majesty's command, published by Carl Linnæus, Stockholm, 1715. What has ultimately become of these valuable collections is unfortunately not known.

To P. ELVIUS.

"SIR,

"I consider with reason that one of the greatest advantages which the Academy possesses, is that the Royal Academy has obtained permission to send a person gratis on behalf of the East India Company to an unknown part of the world; if all other means should possibly ever be debarred, this one alone can maintain the lustre of the Academy.

"Besides, this is just the privilege which can give the Academy lustre and reputation throughout the world; this alone may compel foreigners to learn and read the treatises, and to make them indispensable to men of learning. This advantage has never been possessed by any Academy of Science; this solitary privilege does more than all that the great Louis XIV. expended on Tournefort, Plumier, Trevillée, Lignon, Surian, Lafiteau, whom he sent to the West Indies to botanize, for this will be done honestly. Thus, really, the Royal Academy owes more praise to him who has procured this advantage than ever the Parisian Academy did to Louis XIV.

"May God avert from the Academy, that the

time should ever come, that this priceless opportunity, through recommendation, be given to any unworthy, who should not understand anything, or who under pretext of affecting something, merely sought his own advantage, and private trading on China. It would be a disgrace to the Academy, ingratitude to those who had procured this, and an affront to its members.

"However, I congratulate the Academy with all my heart, and congratulate myself who am allowed to participate in such great delight, I feel new desire to live a few years longer, to witness how our nation shall shine brilliantly amongst all curious nations; we, who before have scarcely known how to tell the difference between fir and larch, shall now be able to teach foreigners to count the eggs in the polypi. I have heard that this winter two ships are to be sent out by the East India Company, one to China, and one to Bengal, which is the more superabundant in everything that is rare; there no one with opened eyes has yet botanized, much less done anything in Zoology; a few blind ones have picked up about fifteen herbs there, which are all somewhat rare. I never could have thought that the Swedish nation should there have got the honour of describing its rare objects. I thank the Academy with due respect, that it has been pleased to hear me regarding this momentous appointment, I vow by all that is honest that I shall never abuse this privilege, but shall always advise the best I know and am acquainted with, without regard to friends, recommendations, or any self-interest, to what I have in common with Science."

In September 1747, Linnæus wrote officially to the President and Members of the Royal Academy of Science, urging Professor Kalm's journey to Canada; an extract from this letter is highly characteristic.

" My respectful request to the Royal Academy is now, that Professor Kalm's journey with all diligence be urged on, for I have from my long experience learnt how necessary it is to strike while the iron is hot, and to work while the cause is young and fresh, while the stars are favourable, while one has yet the wish, hope and longing, while the fancy is tickled, and while one has patronage. If fate gains time

The Floral King. 159

which is always envious of great things, it gains a great deal, one must dread time for every day."

Professor Kalm wrote the 14th October, 1748, from Philadelphia in New Sweden in America :—

"Wild mulberry trees, of several kinds are found in great abundance everywhere in the forests; those who have travelled all the way to the north of New England, where the cold during the winter vies with that in Torneo, and the Lapmark, assure me that they have seen there also wild mulberry trees, not one, but several have for curiosity kept silkworms, which they fed with them, and which have spun as good silk as any that exists in the south of Europe. One of the former Governors of New York, got early from his own silkworms, which were fed with the leaves of these trees—as much silk spun as he needed for his whole family: but because the day labourers here are extremely dear, and because people find their greatest profit in raising corn, which is exported from here to the whole of West India, they have forgotten all about breeding silkworms."

Referring to the intimation which was published in the Scientific Journal, Linnæus wrote to P. Elvius,

"I leave it to you, sir, to judge whether one ought to give one's opinion so publicly about the mulberry trees, who knows if the English may not prevent us and begrudge us this! More likely better to keep silent regarding a great many expectations until one has them safe in hand; regarding which confer with Baron Hårleman."

Linnæus seems justified in his suspicion of the English jealousy, for he wrote in 1755 to Wargentin, "Some years ago an *Oeconomus* in England, sent the Academy of Science a specification of a number of results arrived at, each which he would reveal for a certain stipulated price. Let me know what is his name. Amongst those were also a new kind of grain, which was incomparable for husbandry, and was preferable to wheat, and of which a few grains accompanied, and being present I got a few. I well observed that they were purposely dried so that they might not grow, but one came up and has now increased itself, so that this summer I shall get a small quantity. Let me know all that he wrote respecting this kind of grain, and what he demanded for it. This promises well."

The red vein of the discourse is again met with, and contained in the preface to the later editions of his *Systemæ Naturæ*, the first part of which is kept in the archives of the "Linnæan Society," in London, a draught written in the hand of Linnæus himself, reads as follows :—

"I beheld only the back of the Infinite, Omniscient, and Almighty God, where He went forth, but I felt dazed. I tracked the footsteps over the fields of nature, and I observed in every one—even those which I scarcely could descry—an infinite wisdom and power, an inconceivable perfection; I saw there how all animals were maintained by the vegetation, the vegetation by the soil, the soil by the globe, how the globe was turned night and day around the sun, which gave it life, how the sun with the planets and fixed stars rolled as on an axle, an inconceivable number, and infinite space, and were kept up in the void nothingness by the incomprehensible original motive power, all things' Being, the commander and mainspring of all causes, the Lord and Master of the world. If we wish to call Him Fate, we commit no fault, for everything hangs on His finger;

if we wish to call Him Nature, we neither commit any fault, because from Him everything has originated; if we wish to call Him Providence we also speak rightly, for everything obeys His will and guidance. He is entirely Sense (Sensus); entirely Sight; entirely Hearing; He is Soul (Anima); He is Spirit (Animus); and He alone is self-sufficient! No human guess can comprehend His form; it is enough that He is an eternal and infinite, divine Being, who is neither created nor born; a Being without whom nothing exists, that is made, a Being who has founded and built all this, who everywhere shimmers before our eyes, without our being able to see Him, and who can only be beheld by our thoughts, for such a great Majesty sits upon such a sacred throne, that there no one is admitted but the soul."

The Botanical garden at Upsala, founded by Olof Rudbeck, senior, was perfected by Linnæus to such an extent, that it could vie with the finest gardens in the world, and no doubt surpassed them as regards scientific value. Burser, who wrote and published a description of Upsala in 1773, remarks,

"No garden is better planned, no other surpasses it in the numbers of rare plants, and no other has ever produced so many different kinds of seeds, although it is situated in the coldest climate of any Botanical garden in Europe." Linnæus' house stood adjoining the garden.

Like in the youth of Linnæus, the botanical excursions presented a lively picture on his resuming the duties, for he writes in his diary about himself: when he every summer botanized, he had a couple of hundred auditors, who gathered plants and insects, made observations, shot birds, wrote protocols. And how from 7 o'clock in the morning till 9 o'clock at night, on Wednesdays and Saturdays, they had botanized, returned to town with flower-decked hats, and with kettledrums and bugles, accompanied their leader through the whole town to the garden. Several foreigners and gentlemen from Stockholm participated in these excursions of Linnæus. But then Science had just reached its height.

Professor SÄTHERBERG says :—

"Linnæus' *Systemæ Naturæ* comprises the three kingdoms in nature, the animal world, the vegetable world, and the geological world. In his small treatise: *Deliciæ Naturæ*, this genuine Linnæan painting of nature, so rich in fancy — Linnæus speaks of nature as consisting of three temple-courts, that of Pan, Flora, and Pluto, who each holds sway in his respective dominion."

Linnæus was elected corresponding Member of the French Royal Academy of Science, the 17th of June, 1738, and chosen Associé de l'Académie des Sciences, on the 8th December, 1762, succeeding the great astronomer Bradley on his demise. This nomination was sanctioned by the King on the 11th of December. Contemporaries, foreign Associés with him, and others of the French Academy, were Poleni in Padua, Morgagui in Italy, Bermsulli in Basle, von Swieten in Vienna, von Haller in Sweitz, Euler in Berlin, and Macclesfield in London. To this, considered the greatest honour that can be conferred upon any scientific man, Linnæus was the first of the Swedish nation that had been elected.

. . . .

To P. W. WARGENTIN, 1772.

" It is made known that a new Society of Science

has been instituted in Rotterdam (*Soc. Batavo Roterodamiensis.*)

"They have sent me their *Institutum* which is copious. Many of the principal men in Holland are members. Many Dutch Professors, &c. It seems as if they were in earnest, because the contributions in advance are considerable. They have also done me the honour of making me a member, but I am now getting tired of corresponding with so many for my Societies are already eighteen in number :—

Holmensis,	Cellensis,
Upsaliensis,	Bernensis,
Petropolitana,	Zeelandica,
Berolinensis,	Roterodamensis,
Natur-Curios.,	Londinensis,
Angl. œcon,	Tolosana,
Edinburgensis,	Florentina,
Parisiensis,	Nidrosiensis,
Monspeliensis,	Philadelphica."

. . . .

Patrick Brown was born in Ireland, 1720 (died 1790), studied medicine in Paris and Leyden, in which latter town he made the acquaintance of

Linnæus. Brown practised as physician in the island of Jamaica, where he made great and valuable collections in objects of Natural History, regarding which he published: " Civil and Natural History of Jamaica, and its natural productions, fossils, vegetables and animals." London, 1756. From Linnæus' own diary it is evident that through purchase he became the possessor of this; he also wrote in July 1758, to his friend Abr. Bäck: "At last I have got Dr. Brown's *Herbarium Jamaicense*, which I have expected so long. It consists of 1000 rare plants. I cannot sufficiently express my surprise that any Englishman should allow such an excellent American collection of the rarest plants to leave his country for 100 plåtar. How busy I am now, you, my dear friend, can easily imagine. Here I find all those plants which I had got a glance at in Rolander's collection, and all which our dear departed Löfling described. This gives me so much to do, that I forget friends, relatives, home and country. I am now glad that Salvius did not sooner get paper to the second volume of the *System*."

In Linnæus' autobiographical note of 29th June,

1756, he writes: "Rolander on his homeward journey from Surinam sent a Cactus with Concionelles, in a pot." While Linnæus presided, the gardener took up the plant and cleaned away all dirt, consequently all the grubs, and transplanted it to another pot, so that although the worms had fortunately arrived safely, they perished in the garden before Linnæus saw them, and so vanished all hopes he had of getting them, which he thought could be cultivated with profit in the garden. This moved him so that he got migraine, one of the most dreadful paroxysms he had ever felt.

However two of the little worms were saved alive, to some consolation for him.

"However, in the meantime, one is compelled to receive abuse from stupid people, who themselves dont know *curba*, much less its species, that's the reward here in Sweden, where after the manner of the Germans, we labour to refute that which we do not ourselves understand, and hate what is in any way remarkable, because some one else has first observed it."

Of Linnæus' twelve disciples, who as apostles of

Science, were sent out all over the world, six perished on their perilous journeys, as martyrs to the sacred cause, while the other six were fortunate enough to return with rich collections of the natural treasures, strewn all over the world by a bounteous Creator. Those who returned, and whose names live in conjunction with their more famous teacher, were Kalm, Rolander, Torén, Osbeck, Sparman, and Thunberg, whilst those that perished in the cause, but whose natural collections in most instances reached Linnæus, were Ternström, died 1745 at Polo Candor, one of a group of islands near Cambodscha; Hasselqvist, died 1752 at Bagda, near Smyrna; Löfling, died 1756 in South America; Forskål, died 1763 in Arabia; Falk, died 1774 in Kasan, Russia, and J. J. Björnståhl, died 1779 in Salonica, Macedonia.

CHAPTER XIV.

"NUOVE SCOPERTE *Intorno le Luci Notturne dell' Aqua Marina, Spettanti alla Naturale Storia, fatte du Guiseppe Vianelli, Medico-Fisico in Chioggia, e consecrate a Sua Eccellenza H. N. U. Signor Giralomo Giustiniani. In Venezia appresso Francesco Pitteri* M.DCC.XL.IX. *con licenza de' Superiori.*"

HIS little book which consists only of two sheets octavo, was printed a few months ago, and sent me by post from a nobleman in Venice. As it contains something new, and hitherto unknown to the whole world, through which science has been enriched, I thought I would render my

country-men an acceptable service by making this generally known among us.

"To each and all who have voyaged at sea, it is well known that the water which roars around the ship, and causes the foam through the speed of the vessel, is often luminous as if it were fire, which men learned in Natural History have assigned to various causes. *Physici* have given their opinion that it is caused by an electric power; chemists have thought it comes from the salt and its phosphoric quality, but our author, Signor Doctor Vianelli, is the first who has correctly explained the cause in his work.

"He has observed that the water in the bay at Venice is luminous only from the beginning of the summer until late in the autumn, particularly where 'tång' and sea-weed grow, and mostly where the water is moved by waves, ships, or oars.

"The author took, in 1746, some such luminous sea-water in a vessel and brought it home with him, it was then no longer luminous, but when he splashed it with his hand it immediately became luminous in the dark. By daylight he scrutinized if any extraneous beings were to be seen in the water, which would cause such a light. He could not, with his naked eye, discover anything, he therefore filtered the water through a piece of thick fine cloth, when the cloth became luminous in the dark; but the water was no longer luminous, although he splashed it with his hands. From this he found reason to believe, that what was luminous in the water existed separately from the water itself. On the cloth which was luminous after the filtering, he saw that the light consisted in an infinite number of luminous particles or dots, but as he had then no *microscope* he was compelled to leave them unviewed. Afterwards he got a *microscope*, then gathered seaweed, which mostly is luminous at night time, when he kept that in a dark room he saw more than thirty luminous dots on one leaf. He shook these dots over a paper, when *one dot* fell down on the paper and glowed; it was fine, and the size of half an eyelash, and dark yellow in colour. After this he took the microscope and saw that it was a living grub, which consisted of eleven rings, like most *lava*, and had by its sides even as many pairs of brushes instead of feet, and both at the head and at the tail four threads, *Antennæ* or *Feutacula*. Afterwards he frequently viewed them, and found that all which was luminous in the water or the foam, was caused by these tiny, and almost invisible worms.

"These grubs glow with the whole body, and not like the glow-worms with a part, but not particularly when they lie quiet. The light of these luminous worms keeps during the spring everywhere on the sea-weed, and mostly to the surface. When the water is very luminous during the nights, the fishermen predict that storm and foul weather will arise, which is caused by the worms then being more active and disquiet.

"From these results of Dr. Vianelli, it is incontestible that the foam of the sea is luminous from worms, and also that *Peuna Marina* glows in the dark (about which Mr. Shaw writes that the fishermen at Algeria often get them with their nets, when in the nights it glows so that they can see the nearest fishes in the net) indubitably caused by the little worms. I fain wish that the author had defined these worms, and if I am to believe his eyes,

The Floral King. 173

I cannot picture to myself else than that these worms belong to the Genus *Aphrodita*.

"The author has else set out his little treatise in polished style, with various verses. Furthermore all his results are contained in this rendered account."

. . . .

To P. W. WARGENTIN.
"SIR,

"Kalm's discovery of Lobelia is great, and greater than I dare to relate.

"Hasselqvist is not less an honour to himself. He has described *Cycomarum* with its *stupenda historia*.

"How 200 people during two months lived solely upon *Gummi Arabicum*.

"Regarding a kind of rat or hare, which with its forefeet never touches the ground, always jumps like a grass-hopper.

"Complete description of *Camelespardalis*, which has never been done before.

"A small *Casuarius* or ostrich which grows no bigger than a sparrow.

"*Aspis* correctly described, which it has never been before.

"Regarding *Jaculus* or *Serpens Evæ* who always goes erect.

"Regarding *Gecko*, which blows a dangerous poison through its feet.

"Twelve kinds of new genera of fishes from the river Nile.

"Fifteen new genera of insects.

"The ant, which is one of *Plagis Pharaonis*.

"Regarding the tape-worm in Egypt.

"That which causes the eye-complaints of the Egyptians, regarding a kind of itch, caused by the flooding of the Nile.

"In one word I got so many new things by the last mail from Hasselqvist, that I have never seen such, neither in letter nor in book."

The list referred to is most interesting, wherefore we insert it entirely as done by Linnæus himself.

"List of observations which Hassleqvist made, and already prepared in Egypt, and which by him were intended for Archiater Linnæus:

"1. Notes about the *Tape-worm* in Egypt.

" 2. The cause of the eye-complaints of the Egyptians.

" 3. Description of a kind of itch at the flooding of the Nile.

" 4. *Balsam de Mecca*, its home, the signs by which it is known, its use in the East, adulteration and description of the tree.

" 5. The use of *Mumias* as medicine in Egypt.

" 6. An unexpected use of *Gummi Arabicum* having for two months sustained the life of some hundreds of people.

" 7. *Sal Amoniac's* preparation in Egypt, sent to the Royal Academy of Science.

" 8. *Cassiæ Fistulæ* preparation.

" 9. The use of *grass-hoppers* as food in Egypt.

" 10. The use of the date tree in the economy of the Egyptians.

" 11. The preparation of indigo in Egypt.

" 12. The growing of safflow in Egypt.

" 13. The tending the rice in Egypt.

" 14. *Minosa Arabis Lebbeck*, full description sent to the Royal Society of Science in Upsala.

" 15. Sycomon, entire *Historia Naturalis*.

"17. *Chenopodii*, two new kinds in Egypt.

"18. *Rhamus Arab Nabea* described.

"19. *Chenna*, much used for yellow colour.

"21. A couple of stones of a strange kind.

"22. Description of all Petrifications in the Pyramids in Egypt.

"23. *Strata Terræ* in Egypt.

"25. *Apinior*, two kinds in Egypt: their description.

"26. *Pharaon*, an animal which goes into houses like cats, with all remarkable concerning it.

"27. A kind of rat, with head like a hare's, snout like a pig's snout, body like a rat, tail like a lion; can never touch the earth with its forefeet but jumps like a grass hopper; frequents the mountains between Egypt and Arabia. The whole description of this really wonderful animal is sent to the Royal Society of Science in this place.

"28. *Camelo-Paradalis* which has scarcely ever been seen by any one but Bellonius, its entire description sent to the Royal Society of Science.

"29. A parrot which is the most beautiful of its kind.

"30. A small *Cheradrius* from Alexandria.

"31. The *Ostrich* and its affinity to the 'Snäppa.'

"32. *Casuarius*, a very small kind of Damiata.

"33. A *Pigeon* with straight feathers, standing erect on its back.

"34. A pretty turtle-dove, common in Egypt.

"38. Four kinds of serpents, particularly annotated with their *Scuta abdominalia*, amongst which are *Cerartes Alpini*; or, the genuine *Aspis* and *Jaculus* or *Serpens Evæ*, which have never before been described.

"40. Two *Lizards* in Egypt described.

"41. *Gecko*, which blows a dangerous poison through its feet.

"43. *Echeneis* and *Sardaigne* description.

"55. Twelve fishes from the river Nile, and which constitute as many new genera.

"56. *Dermetes*, which eats dates.

"57. *Cerambyx niloticus*.

"58. A *Butterfly* from the subterranean passages at Alexandria.

" 60. Two singular and new *Genera* of *Insects*.

" 75. Fifteen new *species Insectorum*.

" 76. *Cancer cursor Bellon*; or, a crawfish which runs on land.

" 77. The *Ants* which course on the sand, by the Pyramids of Egypt.

" 78. The tiny *Ant* which exists in the houses in Cairo, and is one of the seven scourges of Pharaoh.

" 79. The African *Scorpion*.

" All these, and much more have already been described with all particulars in Egypt, by Herr Hasselqvist."

Fabricus, one of his many foreign disciples, writes: " During two whole years I was fortunate enough to enjoy the teaching, guidance, and intimate friendship of Linnæus. When I made his acquaintance he had not yet reached his sixtieth year, but his brow was already furrowed. His countenance was open, almost always gladsome, and resembled much his portrait in *Species Plantarum*. But his eyes—of all eyes I ever have seen, they were the most beautiful. Certainly they were not large, but were illumined by an inner fire, and possessed a penetrating power, which I never have met with to such a degree in any other person. His soul was exalted and noble, although I am well aware that some people have tried to accuse him of various faults. His greatest superiority consisted in the systematic order in which he arranged his thoughts. Whatever he did or said bore the impress of order, truth and regularity. He was excellent good company, agreeable in conversation, and full of funny anecdotes. Easily incited to anger, he became in such moments hasty and vociferous, but his anger was readily assuaged, and he was soon in brilliant good temper again. His friendship was reliable and unchangeable, generally based on mutual esteem in Science.

" Not a day passed in which I did not see him; either I was present at his lectures, or I frequently passed several hours with him in friendly conversation. We were three, Kuhn, Zoega, and myself, all foreigners, and this was one reason he was so particularly friendly with us. In winter time, we lived quite opposite his house, and he came to us almost daily, in his short red dressing-gown, with a green leather cap on his head, and with his long tobacco pipe in his hand. He only intended to stop for half an hour with us, but it frequently happened that he stopped a whole hour and sometimes two. His conversation during these hours of relaxation, was particularly thoughtful and pleasant. He would either be telling us about learned men, with whom he had made acquaintance abroad, or he was elucidating some dim passages, and giving us instruction of some kind or other. He laughed heartily, and his whole countenance gave expression to the good humour and friendliness of his mind.

"In the summer we accompanied him out in the country. Our time then passed still more pleasantly. He lived at a peasant homestead, about an English mile and a half from Hammarby; Linnæus rose very early, generally at 4 o'clock. About 6 o'clock he came to us, for his house was being additionally built to at this time, he breakfasted with us, and afterwards instructed us about the natural orders of plants. We were generally occupied with this till about 10 o'clock, and then we went to some rocks in the neighbourhood, the examination of which gave us much pleasure. The afternoons we spent in his garden, and in the evening we generally played a game at cards called 'Frisette,' together with the ladies.

"Sometimes the whole family came to pass a day with us, and we then sent for a peasant, who played an instrument resembling a violin, to the strains of which we danced in the barn of our homestead, and although the gathering was but small, and the dance very countrified, the merriment was general. Linnæus sat and looked on, smoking his pipe whilst we danced. Sometimes, but very seldom, he danced a 'polska,' a country ring-dance, in which he excelled all us young men. He was much gratified to see us really enjoying ourselves, even if in that we became boisterous, he was only anxious that we should do so as much as possible. These days and hours will never be effaced from my memory, and to dwell on them affords me much heartfelt pleasure."

Linnæus has been blamed for having in particular favoured his foreign disciples, but there was no real cause for any such blame, for he was equally kindly disposed to scatter the treasures of his great learning to all who thirsted for knowledge, but just because of his genial disposition he thought it his duty to look with parental care to the young strangers who, alone for his sake, came from far off lands specially to benefit by his training at the University of Upsala. And what could be nobler than his renouncing all claim of remuneration from them if their means were small? To one of these, on offering him money, he said, "Tell me, candidly, are you rich, and can you well afford this? Can you do without this money when you will have to return to Germany? If so, leave the money with my wife, but if you are poor, Heaven prevent me from taking a single stiver from you!" To another of his disciples he said, "You are the only Swiss who has ever come to me, and it is a pleasure to teach you gratis all that I know." This generosity did not arise from vanity, nor from depreciation of money; Linnæus had suffered too much real want, not to be careful of money, but avarice was far from his nature, he liked to be saving when with good conscience he could do that, but he loved better still to help where assistance was needed.

Von Haller in Göttingen charged Linnæus with presumptuous pride. "This man," he said, "considers himself to be the second Adam, and gives all animals names, each according to its kind, without caring the slightest what his predecessors have done in that respect."

When busying himself in his beloved botanical garden Linnæus said, "he felt happier than a King of Persia." And great must have been his joy when all the treasures of Natural History came pouring in from many distant lands, as tributes to his genius and learning; for to Upsala came birds and insects, monkeys and fishes from the tropics, while Siberia sent specimens of its furry inhabitants, and the arctic seas of their whales and seals. He must have enjoyed to the full to see himself so loved and admired by all, and the gratification of having created such a general interest in Natural History, that almost everyone had become enamoured of the subject, and ladies, as well as kings and queens, and learned professors, began to form collections of their own. It must also have gratified him to see the Queen Lovisa Ulrica, herself an acknowledged great and brilliant genius, listen spellbound to his lectures, and then with a gracious smile invite him to a conversation with her upon those subjects which none so well as he knew how to make fascinating. But who can record the toil and zealous exertions by which he acquired his wisdom, and harvested superlative joy? Who can tell how deep the wounds were which were ruthlessly inflicted upon his sensitive heart? He had gained many great victories, and silenced many of his adversaries, but nevertheless were now and then home thrusts made at him, and he became more sensitive to unkindness the more love and admiration he was shown. A harsh, unjust criticism, and false charges or envious slander cut him to the heart, and he brooded over it long. Upon one occasion he was visited by insomnia for two months in consequence of a deep affront. If he felt great joy when his travelling disciples, the intellectual offsprings of his genius, the children of his mind, sent home to him from distant lands scientific treasures to testify their love and fealty, he also grieved all the more bitterly when several of these promising young men found an

The Floral King. 185

early grave on the journeys of scientific research, which, inspired by him, they had undertaken.

It naturally became the duty of Linnæus to systematically arrange the collections of all these, and to edit for publication the manuscripts of those defunct, and this, added to his manifold duties as professor and corresponding member of eighteen learned societies, greatly increased his labours. However he continued to publish many great works of his own in rapid succession, and of one amongst them, the *Fauna Suecicia*, he says "that he had worked on it for fifteen years." Several of his first works required new, revised, and enlarged editions, and were from time to time greatly increased in volume; thus he instances having worked the whole of the year 1758 on the tenth edition of his *Systema Naturæ*, and still he revised two more new and enlarged editions of this voluminous work before his death. Upon several occasions he was by the government sent to various provinces to perform scientific researches, and to report upon these. At one time he attempted by artificial means to produce pearls in mussels, at another period he tried to cultivate the tea plant in Sweden, because, "it is imperative," he said, "to close the door through which all the silver passes out of Europe." Then he was busy analyzing the waters of the mineral springs in Sweden, and was also called to the pleasant task of arranging the Queen's private collections in Natural History.

In course of time many of his young and travelling adepts returned, among whom we may mention Osbeck, Montin, Thunberg, Sparman, Forster, König, Afzelius, and Alstrin.

❈ CHAPTER XV. ❈

LINNÆUS mentions in his diary of 1750, that through hard work, and increase of years, he had become afflicted with rheumatism, which gave him excruciating pain, and kept him to his bed with little hope of his life being saved. He cured himself, however, that time by daily eating a platefull of wild strawberries, and continued afterwards every year to use this simple remedy.

When Lovisa Ulrica, (the reigning Queen) was told that wild strawberries were so beneficial to him, she commanded that these berries should be specially grown at all the Royal country residences, so that in all seasons there could be sent such to Archiater von Linné.

The third of May, 1764, he was suddenly seized with a severe illness, and soon there seemed no hope of his recovery, and his wife and children stood around his bed in mute agony, every moment expecting he would die. The door was then softly opened, and in stepped a man, whom no one ever would have expected to see there. It was his old secret enemy Nils Rosén, against whom the enraged Linnæus, in his young days, had drawn his sword in challenge, and for whose sake he had been compelled to leave the University, and the same man who on his seeking the professorial chair had sought to exclude him from Upsala, where he was jealous of him as a superior rival. Rosén was also now a man of great fame, and as a physician even of greater reputation than Linnæus. Rosén had been made professor and archiater shortly before Linnæus; and when he was awarded his patent of nobility, he took the name of Rosén von Rosentein. Since the two professors had exchanged their Academical chairs, they had certainly not showed any animosity to one another, but they had neither showed any desire of forming any friendship with one another. They could not but acknowledge each other's great merits, but they preferred doing so at a polite distance. What then wanted Rosentein at the deathbed of Linné? He had not been called in as a physician. Had he really come to see the man of superior genius, whom in vain he had tried to obscure, sink into the jaws of death? Heedless what the grieving family might think of his visit, Rosén advanced to the sufferer, and fixing his penetrating eyes upon him, seemed for a few moments to sink within himself in deep contemplation. After awhile, in a decisive manner, he gave new orders of treatment, quite different from that to which Linnæus had been subjected. Accustomed, as he frequently was, to battle victoriously with death, he was now determined to tax his abilities to the uttermost, and his genius had discovered new means by which to overcome the ravaging malady whilst he was gazing into Linneaus' features, to see what traces of lingering life there remained. The confidence and zeal with which he gave his directions allowed no opportunity for hesitation from any one, and they hastened to obey him in everything, and the battle which the great physician fought with grim death for the noble prey, was both long and difficult in the extreme, but he persevered night and day, with undaunted skill, never leaving the bedside of his patient. When Linnæus at last awoke from his long trance, he met the sparkling eyes of his old foe, but he was too weak to comprehend the situation, and he looked again feebly, and inquiringly turned to his wife. But when she, with tears of joy, told him that, next to God, they had to thank Rosén for his life being spared, Linnæus held out to him his tremulous hand, greatly affected, and that solemn moment expressed more gratitude than words could convey. From that day they became firm friends for the rest of their lives. Great was Linnæus' gratitude and admiration for his new found friend, who so long had been his rival, and great must have been Rosén's satisfaction to be able thus nobly to atone for the deep wrongs he had formerly done Linnæus.

Convalescent, Linnæus removed to his Hammarby

to benefit by the influence of the country air. Shortly afterwards, he and his wife celebrated the 25th anniversary of their marriage, which also in Sweden is called a silver-wedding, and is a festival particularly observed by all old friends of the house.

At this same event was also celebrated the marriage of his eldest daughter.

Before the end of the summer he was so far restored that he could resume his usual avocations, but his accustomed health and vigour he never regained; and in 1770 he was again dangerously ill, as his diary relates, but at last was again restored.

In his diary, Linnæus has given the following brief characteristic description of himself: " Linnæus was not tall, nor little, thin, brown eyed, lithe, quick, walked fast, did everything promptly, could not bear tardy people, was sensitive, easily affected, worked continuously, and could not spare himself. He appreciated good eating and drinking, but was never extravagant in either. He did not care much for his appearance, but always thought that the man ought to adorn the dress, not *vicé versa*. Consistorial business was neither his pleasure, nor his work, for he was made for, and thought of other things, than what there is treated of and decided upon."

Of his friend Archiater Bäck he says that, "whenever he went to Stockholm, he always stayed at his house, as if it had been his real brother;" and this gentleman has drawn his portrait somewhat more minutely. "Amongst other traits," he relates, "Linné liked good company of an evening, when he was in excellent spirits, jested and laughed heartily. Easily moved to joy or sorrow, or anger, but just as easily assuaged. His heart was excellent to the core, his lips spoke the language of truth and virtue. He was faithful and considerate to his friends, and did not pay his enemies back in their own coin, yet he did not easily forget, saying that he did not like to be deceived a second time. The management of his house he left to his wife and was quite contented to see it in such able hands. A faithful and good husband, no less a tender father, delighted in entertaining his friends well, but as regards his own person careful of expenses; liberal for his science, also when he met any wretched mother with a little child. He renounced what was his due from any poor students, and to the best of his ability provided for the children of his rural district. He revered religion, and did not try to fathom its mysteries. It is said that Boyle and Newton bowed their white heads every time the name of God was mentioned, and so also has Linnæus, on every page of his writings acknowledged the glory of God, and prostrated himself in the dust, exclaiming, 'It is the finger of God! Learn to know the Creator by His works, contemplate His wonders and adore Him.'

"His heart was burning with enthusiasm, and his speech alone on that theme eloquent. It was not his habit by surreptitious machinations to hurt any one, nor could he brook that anyone about him was insulted.

"His genius particularly applying itself to results and experiences, and on that basis founding his knowledge, he has written down a great many things, which had occurred in his lifetime, under title of '*Nemesis divina,* or God's punishments,' to prove the Aphorism, that God also in this life punishes evil doers, a moral axiom which in particular he was wont to impress upon the minds of the youthful students."

It is in "*Nemesis divina*" that Linnæus records that everything went wrong with him as long as he intended to revenge himself; but that afterwards, when he changed his heart and left all in the hands of God, everything prospered with him. Those that accused Linnæus of conceit wronged him exceedingly, for in this book he has written, "No one is the architect of his own fortune," and again, "all that we possess is a loan from God. We bring nothing with us in this world, and we take nothing away with us. When God takes anything back, through fate, which is His Executor, we grieve that we have lost our possessions which were not ours, but merely a loan."

Regarding Linnæus' power of influence over the young students, Hedin relates; "The unrestrained mirth, the sprightly joy, and that frequent uncontrollable impatience, which so often manifest themselves in youth, and cannot brook restriction, never made themselves apparent amongst Linnæus' auditors. No thoughtless fool stretched himself carelessly, or yawned, no wit-snapper turned his words into ridicule, and no censor whispered epigrams.

The Floral King.

When Linnæus was speaking of the Majesty and wondrous works of the Creator, awe and admiration were depicted in all faces, and the most self conceited youths, who always are ready, at the expense of the most sacred truth, to vent their curt remarks, would not have indulged in the vaguest doubt. Did he speak again relative to the prescriptions of diet, he would often allow his pupils to laugh exceedingly, as he depicted the mad whims of fashion, and during elegant and agreeable pleasantry, he taught the most useful precepts regarding the acquiring and preservation of health."

As a proof of the great enthusiasm with which all the students listened to Linnæus, Hedin relates the following in reference to the last public oration which Linnæus delivered when he resigned his Rectorship. " No one was absent of all the students of the University, the *confrères* of each province assembling together, and of all the youths, numbering some 600, scarcely one moved during the whole ceremony. Great admiration, common to all, held all these youths spellbound, who else were such unrestrained and impatient beings. From every eye beamed rapture and admiration. A vociferous *vivat* accompanied the speaker all the way to his own gate, and delegates from the students of all the divers provincial nationalities waited upon Linnæus the following day in his reception room, sent to express the concurring reverence of their respective *confrères*, and on their behalf to request that the oration, which Linnæus, the previous day, had delivered in Latin, might be translated into Swedish, and be printed at their expense."

Linnæus refers himself to this his third and last Rectorship of the University, thus, "during that period, no student had stood accused, no one had gambled, no one wore masks, no disturbance was reported, never before had any term been so quiet."

Narrow-minded people have charged Linnæus with being vain and conceited, on account of the self-laudation which appears in his oft-quoted diary. But all the favourable expressions he uses are irrefragable truths, and he gives utterance to it so naïvely and simply, that this is the best refutation against these charges, besides, a transcendent genius must not be measured by the estimates of common men, for what would be puerile boasts, in the diary of any common mind, only bespeak the childish simplicity and frankness of the man of genius, such as Linnæus.

"Here I have written my own panygeric," he writes to Bishop Menander, when he sent him his diary, by the aid of which he was to write a biography of Linnæus: "I should never have shown it to anyone in the world, if not to the only one of my friends who has ever remained unchangable in his friendship from the time I found myself in less fortunate circumstances. If you, my dear friend, should think it meet to make any extracts from it, it would attract attention when it comes from such a pen as yours. I really feel ashamed to show it to you, and should never have done so if I had not been convinced of your friendship, and unchangeable affection. Dear friend, remould it any way you please, since it is only intended to represent facts."

These annotations were thus never, in their original form, intended to be read by anyone else, much less by the public. Besides they bear testimony to his humility, and express a joyful consciousness of his own worth. Linnæus seemed often astonished at the many honours conferred upon him, and no one more freely acknowledged the merits in others, with regard to the sciences as in everything else. In one place he says, "If Tabucius comes to me with an insect, or Zaega with a moss, I doff my hat to them, saying, 'in these you are my teachers.'"

The summer vacations he generally passed at Hammarby, whither he had removed the best portions of his collections, and stored them in the little house he called his *Museo*. His most beloved disciples gathered here during the summer, and to this rural temple of the High Priest of nature pilgrims came from many lands, and all who had been there spoke with the greatest enthusiasm about Linnæus, for a heaven begotten sympathy seemed to exist between the teacher, his surroundings, and the creed of harmony and order which he taught. At the peasant homesteads in the neighbourhood, lodged the many disciples of Linnæus. Accessible to all who sought him to acquire knowledge of Natural History, he had a marvellous capacity, even when advanced late in years, to find time for everything and everyone.

In his diary for 1771, for instance, he mentions that he taught foreign students eight hours a day in his museum, and he, by no means, confined himself, only to this work.

Year 1774, Linnæus had, what he himself calls, his first warning of death, for he had a fit of apoplexy while lecturing in the Botanical Gardens, on the 3rd of May of that year. His subsequent illness was also, this time, overcome, although very slowly, and first at Christmas time he had mostly regained his wonted strength. The main instrumentality was a valuable gift, which, more potent than any medicinal remedy, influenced him for the better. The King, Gustavus III., had then sent him a collection of herbs and natural history objects, which had just arrived from Surinam, consisting of four cart loads of plants gathered with blossoms and fruit, and carefully preserved in several hogsheads of *spiritu vini*. He got, as it were, new life from the intense pleasure which he experienced in eagerly arranging and describing these rare plants during Christmas-time, forgetting all about his ailings, and in 1775 he felt himself quite hale and sound.

Gustavus III., himself a man of great genius, vastly admired Linnæus, and while yet Crown Prince of Sweden, he paid a visit to Linnæus' museum at Hammarby, and when he had become King he once came to Upsala on purpose, solely to see Linnæus.

In 1776, a great sensation was caused at the University of Upsala, because from Stockholm came the news that the Faculty of Medicine would be deprived of the right of promoting candidates to Doctors of Medicine, that this honour would instead be conferred upon the Committee of Medicine in the Capital. To ward off such an humiliating blow from the University, it was thought expedient that the venerated Linnæus himself should repair to the young Monarch, and plead the cause of the Faculty. Linnæus was, at this time, suffering from illness, and in a very weak condition, so that he could not undertake the journey without considerable difficulty, but still, he would not refuse to go. He did not hesitate to sacrifice himself for the sake of the beloved University, and he left Upsala in the company of Professor Sidrén for Drottningholm, the country palace at which Gustavus III. was staying at the time. A private audience was at once granted to the honoured professor, and as soon as the doors to the royal apartments were thrown open from the ante-chamber where Linnæus was waiting, he tottered towards the King.

"It will never do, your Majesty," he began, forgetting all about the etiquette of the Court, in his anxiety for the honour of the seat of learning. "It will ruin the University and Science; I shan't survive this calamity." Gustavus evidently surprised, did not understand for the moment what was the matter, and turned inquiringly to Sidrén, who, in a few words explained the subject to the King, who then, smiling, patted Linnæus on the shoulder in a friendly manner, saying, "It shan't happen, my dear Linné, you may rest assured, you can return home without fear." And with that the anxious question was settled for the best. Linnæus had one more supplication this same year to make to the King, and it looked at first as if it had not been graciously accorded. Linnæus' only living son, Carl, had already, fifteen years previously, been appointed successor to his father as Professor of Botany. He was only twenty-two years of age when, for the sake of his father's merits, he was promised the chair of Botany after Linnæus, and at eighteen he had been nominated *Botanices Demonstrator*, an office which was then created. Well nigh worn out, Linnæus was much in need of rest, and he asked now for his discharge, with permission to leave the Professorship to his son. "No!" answered the King hurriedly and with determination, "I will grant with pleasure anything you may ask, Linné, but no resignation." With manifest displeasure the supplicant withdrew, for he felt himself hurt, and did not care to hide his feelings. . . . But what appeared a humiliation was not meant to be so. The professorship was in every regard, except in name, confided to Carl von Linné, junior, and the King declared that he, with pleasure, conferred double salary, with the only provision, that the far-famed name of Archiater Linnæus still should be registered as one of the teachers of the University, as if he were still exercising his professorial duties.

Gustavus was fond of brilliant display, and he

could not bear to think that the University of Upsala should not still boast the lustrous name of Linnæus, senior, as if in the exercise of his great calling. Linnæus lacked nothing in reciprocal compliments to the King, for by way of gratitude for many honours bestowed, he nominated the grandest and most beautiful tree he could find in the entire world after this his well-beloved King, *Gustavia Augusta*.

Towards the end of 1776, Linnæus' health was rapidly failing, and his life's lamp flickered as if ready to go out. That spring term he had still lectured, but during the latter part of the year, both his mental faculties and bodily strength rapidly declined.

"Linné limps, can scarcely walk, speaks incoherently, can barely write," those are the last words he, with tremulous hands, has written in his diary. Soon he could no more move from the place where he was seated or lay, could not dress himself, nor even feed himself, but had to be tended like a child. His consciousness was obscured, he gradually forgot everything, even his own name.

But sometimes moments occurred when it seemed

as if the shadows of night, which were setting in upon him, were lit up by the sunset of life. That happened when some one of his beloved disciples called to see him, and when he heard something spoken about any objects of Natural History, or when his friends placed before him some work on Botany, a shimmer of joy then spread over his wan countenance, and a flash of his former genius seemed for a moment to illume his face. But this transient gleam of reminiscence passed quickly away, and left him more prostrate than before.

Linnæus passed his last summer at Hammarby in 1777. As often as the weather would permit he was carried out in the open air, amongst the flowers of his garden, or up to his museum, and his health visibly improved. The beauty of nature was the best balm for him. But his time was nearly gone, and although after his return to Upsala in the autumn, he was so far recovered that he could take a drive every day and sit smoking his pipe, still there was no hope of his ultimate restoration. Tired of life, and forgetful of all his honours, he breathed his last, peacefully, on the 10th of January,

1778. His remains were interred in the Cathedral of Upsala, whither they, in solemn silence, were followed by all the teachers and students of the University. His sorrowing widow had an unostentatious stone put over the grave, but afterwards a monument of black porphyry was erected, to record to coming generations where rest the dust of one of one of the noblest sons of Sweden. The grief at his death was as general and sincere as the admiration and love had been for him while he lived. His works, and his name, live for ever.

CHAPTER XVI.

EXTRACT FROM A LETTER OF LINNÆUS TO HIS AGED MOTHER-IN-LAW—ELIZABETH MORÆA.

"THAT you, dear Madam, are getting tired of a large housekeeping in advanced years, and dear times, is not to be wondered at. I have known many who have got tired of laborious occupations and retired to rest, but from that cause it has happened alike to all of them. When the body no longer gets exercise, having been accustomed to work, it soon becomes ailing, and when the mind no longer has cares and activity, it settles into melancholy. This would, for certain, shorten your days, dear Madam, more than the worst drudgery.

"If you, dear Madam, sell your landed estates and retire, the consequence will be that you, dear Madam, will be cheated of your money and get an ailing body; the home will be broken up, the children at variance, the love diminished, and everything ruined.

"Nothing is so stable that it cannot be destroyed in the morning, and the strong die as readily as the weak, whereby expectations of long forestalling are wrecked.

"Parents frequently love their children more than themselves, but the children do not return the love, if interest be not the attraction, for the branch takes its nourishment from the stem, but the stem does not get it returned from the branch. Look at others who have taken up their abode with their children, and given them everything, if that ever has produced that emulation in respect towards the parents, as when they get the inheritance after their demise.

"You, dear Madam, who have so long loved that

man of honour, your dear and kind husband, my respected father-in-law, defunct, pray remember that he were dead, if he left no children behind him. Our brother Petrus is, as regards his body, as surely a part of you, dear Madam, as your own hand is. His life is my dear father-in-law's as surely as the arm belongs to the body. Brother Petrus is thus my father-in-law with you, dear Madam, or a part of both. My deceased father-in-law lives and continues in the children as long as the children and grand-children live, and increase. Or can one sever a branch from the stem of a tree, and say that the branch is not as much the tree itself as the rest of it? You, dear Madam, favour the increase in your branches, the best you can without ruining the stem, because then the whole tree will die away. I am not particularly acquainted with brother Petrus, but least of all with him of all my brothers and sisters-in-law. That I love brother Petrus is nothing else but that I love the stem. If this perishes, then there is no more relation, no reliance between relatives, against the natural law of humanity. This is the cause of marriage, of love, of everything. If I may be more circumstantial and quote an example, the tape-worm (here followed a figure of the tape-worm, sketched by the letter-writer.) Each part of the worm has its own life. If these joints be severed, still each joint lives and increases. Which part then is the more than any other part of the worm? The last joint produces its offspring, that one, one in its turn, and so ever on. There is no more difference in other animals, than that each joint immediately falls off from one another, but if the navel-string became continuous the children would be strung together like tape-worms, and then would it appear more evident that they were one. Brother Petrus, and the other children, are thus nothing else than yourself, dear Madam, with my dear father-in-law, defunct. God bless the venerable old tree, which has flourished so well on the estate of Sweden; may He allow it to remain constant until times remote, and rooted in blessings, that its branches on all sides may spread to an ornament for the country, and that no branch may wither, for that would immediately make the tree deformed and unseemly, it may still recover itself in time, alone the stem remains firm, and the roots are not damaged.

"I hope that you, dear Madam, will read this as kindly, as I hastily and honestly have written it.

"I remain with unalterable veneration,

"My dearest and kindest Madam,

"Your most obedient servant,

"CARL LINNÆUS.

"Upsala, 1758, the 16th October.

"My wife begs me not to send this letter, for she says she will then never get another letter from her dear mother. But I cannot see how it is possible that you, dear Madam, would take ill what is not meant for offence."

However his prudent wife persisted in her opinion as to the dangerous tenour of the letter, admonishing the old lady not to sell her property, thereby to favour any one of her children in particular, and it was never sent, but Linnæus' wife kept it secure under lock and key in her own drawers, hence years afterwards it was first brought forward among other relics of Carl Linnæus.

The Floral King.

"MY DEAR RELATIVES,

"MY BROTHERS-IN-LAW, MY SISTERS AND MY BROTHER.

"It generally happens, that the fledgelings which have been hatched in the same nest, as soon as they are feathered fly away, each one in his own direction, so that they seldom twitter together on the same tree.

"Fate has been propitious to you, in that it has ordained that you should live and dwell together on the soil of your forefathers, when I have been thrown so far away from my dear relatives, where I must live a stranger by myself.

. . . .

"From my youth trusting in my God, I have worked and toiled, yea, it may almost be said, more than anyone else, and have experienced the truth that the more a lamp burns, the sooner it will be consumed; also that dry log-fuel never lasts so long as the damp wood. The Supreme Being has granted me His blessing in my calling, and given me all that my heart has asked and wished for. I have got a respected appointment, a careful wife, dear children, and some little property.

"I have been made Doctor, Professor, Archiater, Knight, and Nobleman.

"I have been allowed to see more of the wonders of Creation, in which I have found my greatest delight, more than any mortal that has lived before me.

"I have had my apostles sent to all parts of the world. I have written more than anyone, who is now alive: seventy-two books of my own hand are ranged on my desk.

"I have now a great name, even as far as the Indies, and have been acknowledged as the greatest man in my science.

"I have become a member of most societies; of Upsala, Stockholm, St. Petersburg, Berlin, Vienna, London, Montpellier, Tolosa, Florenz, and now recently of Paris, with salary among the eight famous men in the world.

"But when a tree has reached its height, it must fall, for *quidquid ad apicem pervenit, at exitum properat.*

"I have now for the last year observed how age hastens on, for: 'In the day when the keepers of the house shall tremble, and the strong men shall bow themselves, and the grinders shall cease because they are few, and those that look out of the windows be darkened . . . . Or ever the silver cord be loosed, or the golden bowl be broken, or the pitcher be broken at the fountain, or the wheel broken at the cistern' [Eccles. xii. 3 & 6]. I have therefore begun to look to my house.

"It is no less art to turn about in the haven, than to sail out on the boundless main with filled sails.

"I have, therefore, during the last year, begun to build additionally at my country estate, that my widow and helpless children may have a roof to shelter them when I depart from them.

"I have made my disposition as regards my property for my children, and disinherited my son, because he has his salary as Adjunctus to solace himself with (4,000 dal.) until he gets more. And now I have at last sent in my petition to His Royal Majesty for a gracious discharge from the professorship, which I hope to get very soon. I require also to rest, and breathe freely, a day before I die; yet, I have reserved to myself:

"1. To retain my salary until my dying hour;

"2. That my son may become authorized to the professorship.

"3. That I may occupy the chair as long as I please, while he perfects himself furthermore.

"To my son I have only given my library of books, manuscripts, herbarium, naturalies, to a value of 70,000 dal., but for my four helpless girls I am concerned at heart; I have portioned to them the remaining shreds of property, real and moveables, which I have been able to scrape together in this world. I have my pleasure now on Sundays to go to my country seat, to free myself from town-bustle; when I am there I think a hundred times of my absent brothers and sisters, and wish they would pay me a visit, while I still remain in this world: but my wish seems to be in vain. I have plenty of visits from plausible friends, but not from those I most desire.

"I well understand that the days of my dear relatives by degrees hasten to their end.

"May the Supreme Being ease the ailings of old age, that at last we may depart this world, and praise Him for the time He has granted us. I commend myself to the intimate friendships of my dear relatives, and remain,

    " My dear friends,

      " Your faithful brother,

        "CARL LINNÆUS.

"Upsala, 22nd Mardii, 1763."

. . . .

  " MY DEAR BROTHER AND SISTERS,

 " BROTHER-IN-LAW AND SISTER-IN-LAW,

"THE CHILDREN OF MY SISTERS, AND BROTHER,
    AND THEIR HUSBANDS.

" When I now sit on Christmas Day in quietness I cannot refrain from thinking of the dear place of my birth.

 Nescio qua natale solum dulcedine cunctos
 Ducit et immemores non sinit esse sui.

" I have there left many, yes, all my dear relatives who there are allowed to dwell together in intimate friendship; when on the contrary, here I am *peri-grinus in patria*, and have no relatives or friends, when all my *collegæ* have relations. God in His grace has given me my livelihood in a strange place, *sed nihil ab omni parte beatum.*

"When I now contemplate how incessant work, during many years, has made me emaciated, grey, and bent my back, I find full well how time speeds, and that I never more shall see my several dear relatives, which most of all touches me at heart; when my dear brothers and sisters are allowed to see each other as often as they please, you are allowed to comfort each other in adversity, and rejoice at each other's fortune in prosperity. I have no confidential friend, where all my confidence is expelled, he who compliments in prosperity, helps to push the wheel of misfortune down the hill in adversity. You will once, when tired of the world, be gathered to your forefathers, and rest your bones in your ancestral grave, where mine will never be allowed to come. You are allowed to live and dwell in the homes of your birth, and to live with a people brought up in simple honesty itself; but I must always be on my guard against artfulness and schemes. I never forget your innocent Christmas games, but am daily reminded of the wiles of the deceitful world.

"You are allowed at Christmas time to talk in simple confidence, and to jest with one another, at the time when we have spread the table for the most ferocious beasts of prey. My wife and family go to Fahlun at the New Year, scarcely twenty miles (130 English miles), to see their relatives. I remain here alone because I cannot go fifty to see mine; but everyone must be contented with his lot in this world.

"If God grant us health, we have more than *Crœsus*, *Solomon* and *Alexander Magnus*, if we know and understand it. We all begin now, brothers and sisters, to become aged. I am oldest, age has broken me the most. Brother Samuel is still said to be at his best vigour, but now first enters into the years of an old man. Brother Höök and Sister Anna Maria, are likely broken by age and cares. I pity Sister Sophia Juliana, who must sit like a lone bird on the roof, since the nest is destroyed. Our sweet Emerentia, dead, was first called away from the distressing life of this troublesome world. God Himself has been her daughter's guardian. When now we all soon must depart, let us provide well for our children, for afterwards we cannot help them. The tears of our parents for their children rose through the clouds, and stopped not before they were before God's countenance, who looked to it, and built the houses of their children. That same God liveth for ever; happy our children if He vouchsafe to become their guardian. He alone saves the children of the poor, when those of the rich often do not get a morsel of bread to put into their mouths.

"Up here everyone complains of hard times, iron masters and mine owners cry out as loudly as they can that they are being undone. The rich *cedera bonis*, as fast as they can. The country people complain to-day when last year they got fifty-four dlr. for a tun of grain, now twenty-seven, and no corn is in demand. Credit is lost, and one cannot help another, money seems to have vanished, the poor alone seem to find no change. We have had a wetter autumn this year than I ever remember. The winter-rye, which was early sown, is all rotten, the poor who did not possess any rye to sow before they had threshed, have their fields preserved and verdant. Here is spoken of the urgency of the

Diet being called together, but as yet nothing is decided, God grant it may pass off well when it does take place.

"Their Majesties were here at Michaelmas, when they vouchsafed their usual grace to me. The number of students yearly decreases, considerably at the Universities; if hitherto there has been too many clerical germs, soon there will surely be too few if this continues. I send my greeting to all my good, kind, honest parishioners, but there will only be few who remember me, I think I see a new world in Stenbrohult, and all those I knew in my youth are gone to rest. Thus time changes everything, and rolls on like the billows of the sea, of which there is no track left behind, our life is like the corn in the field, which germinates, grows up ripens and is cut off, next year comes new corn from the harvest of the previous year.

"May Almighty God, during the new year now entered upon, bless Brother Höök and Sister Anna Maria's house and family, Sister Sophia Juliana and her children and son-in-law, Brother Samuel and his Mistress and children, and our dear dead Sister

Emerentia's daughter and her husband, and if there are any more of the family.

"To them all I send mine, my wife's and children's and also grandchildren's loving affection, and we will, with united voice, call down the blessing of Heaven upon you.

"May God, during this new year, avert all sorrowful occurrences, and that if we cannot see each other, with joy we may hear of each other's welfare. I live and die,

"My dear friends,

"Your humble and faithful servant,

"CARL V. LINNÉ."

This letter, a touching expression of patriarchal affection of the venerable sage, was rescued from oblivion in 1849, when it was met with in the possession of an old peasant in Småland, who in his youth had been a servant of Rector Höök. This curious old relic, delapidated and worn, has since been photographed, and the fac-simile distributed, and has also been inserted in various Swedish periodicals.

In 1758 Linnæus bought the estate called

The Floral King. 221

Hammarby, situated about six English miles southeast of Upsala, on the great plain surrounding that town. It then had only two small, one-storied, log-houses standing opposite to each other. Four years afterwards, in 1762, the new owner had a larger two-storied house, also of wood, built between the two smaller ones, and in this house, which is still retained in its original form, are kept various belongings, with which in Linnæus' time his house was furnished, and bought for £1,666 by the State in 1879, to be kept as a monument to Linnæus. On the space in front he had planted two chestnut trees and two Siberian appletrees, of which one of each yet remains.

Down in the garden, near the rustic wooden enclosure, a little to the left on passing from the dwelling house, is to this day a square place, surrounded by tall trees, which was pointed out by his grand-daughter as "grandfather's arbour," and here he used to sit and smoke his long pipe, and where also the dinner used to be served, as often as the weather would permit. Many plants which now grow wild there in great abundance, such as *Eronymus europheus*, *Myrrhis ordorata*, *Tulipa sylvestris*, and a few shrubs and trees, are lingering, living mementoes of the loving hand that planted their lineal ancestors.

Linnæus had put up in the neighbouring trees some glass bells which chimed melodiously when the wind moved the branches of the trees.

At the foot of a neighbouring hill was "Linné's Grove," and on the hill his "Museum," built in 1769, now surrounded by a tall fir and pine forest, but the hill was then denuded of trees and afforded an extensive view over the fertile plains, out upon which he loved to look.

In the little house built of brick were kept his Herbariums, the greatest in the world, his *Zoophylæ Conchylia*, insects and minerals, which were all visited by the curious, many of whom came from distant lands.

CHAPTER XVII.

"May, 1764.

"MY LAST WILL.

Y Son has got the Professorship, which, if it is well managed, gives interest on 100,000 dal. copper coin. The Herbarium is worth 50,000 dal. the insects 10,000 dal. the amphibies 10,000 dal. the stones 10,000 dal. consequently he will get no portion of any real estate. My yeomanry farm my widow will enjoy as long as she remains such, and after the yeomanry expenses being defrayed, she shall give to the married daughters their allotted income thereof. But, if to her misfortune, she

enters into new wedlock, which I can well foresee, she will get no share in them, which I bought with the money I earned through night work on books, on Collegiate labours, and by botanizing, and which alone I have acquired. Instead she shall then enjoy her own town-fields, and according to law, a part in all furniture, gold, silver and monies, and that which she inherited from her father and mother, invested in our home, this will reach a considerable sum. This I have written while I still have a clear brain, and this is my last will, over which He holds His hand who always remaineth.

"*Carl von Linné,*

" Witnesses ;—

"ANDERS NEANDER,
"*Phil. Mag. Reg. Ac. Upsal. Ærar.*

"GABRIEL ELMGREN,
"*Phil. Mag. r. and Med. Licent.*"

CAROLI LINNAI TESTAMENTTE.

"Since my years increase, and my strength decreases, I am reminded to look to my house, and keep myself in readiness to bid farewell, wherefore

Linnæus' Study at Hammarby.

I here give my last will, and dispose of my residue as follows :—

"My Son, Professor Carl Linné, who, for the sake of my humble merit, has got from His Royal Majesty, with the consent of the Diet of the Realm, the solace of the professorship I have filled, must not participate in any of my real estate, nor be reckoned any equivalent, but he gets my library, which if he uses together with the professorship as assiduously as I have done, he has far more to content himself than all his sisters together.

" But as regards the movables in the house, in that he will share equally with his sisters, but not double.

" My four daughters get all houses and lands, even Hubu to part between them.

"My wife, provided she does not enter into new wedlock, or dispose of my bequeathed real estate, will reside, during her lifetime, at Hammarby, retains the town-fields, the ground plot with the garden, and enjoys all the landed property, yet in the manner that the married daughters yearly receive from her what Säfja gives above the yeomanry expenditure, that she may have assistance from the farmers at the tillage of the main-homestead soil.

"But if my wife should contract a new marriage, she ought to give over to the daughters all the yeomanry-farms, and of the real estate keep only the field in the town of Upsala, and the freehold estate Edby, with the plot of ground in the town.

"My herbarium, which is the largest the world has ever seen, is to be sold to the highest bidder for the benefit of my daughters, together with the library belonging to it, which is also kept at the *Museo* at Hammarby, yet I would much prefer that the Upsala University would purchase it, for the time is not likely to come when it could again acquire such a collection.

"The other natural collections, such as conchylies, insects, and stones, my son is allowed to keep.

"This after well considered thought whilst yet in possession of my health I have put down and ordained at Hammarby.

"CARL LINNÉ," "SARA ELISABET LINNÉ,"
(His seal of Nobility).   (The seal of nobility).

"That the above written document is made in full

possession of a sound mind, and of a free will, witness the undersigned :—

"JOHAN FLODERUS, Upsala, the 20th August, 1776.

"JOHN HAGEMAN,"

(Seal).

"With reservation of surrogate of my dear mother's sole, inherited, real estate, I acknowledge this testament which I declare.

"Upsala, the 15th October, 1778,

"CARL VON LINNÉ."

In the Autumn of 1758, Linnæus bought Hammarby estate, situated in the parish of Danmark, Vaksala district, province of Upland. Regarding this purchase he wrote to his intimate friend Abram Bäck on the 22nd. December, 1758. " Now dear friend, I am ready to be hung, I have always been in fear of debt, as if it were a batch of serpents, but now I must dance on chatelet which I had never thought. This makes me either prematurely grey, or dishonoured as if it were an ill begotten offspring. For the sake of my little children, I bought, this autumn, a small allodial estate near Upsala for 40,000 dal. A few days ago a whole neighbouring hamlet, five entire freehold farms and leasehold tenements were vacant, or to be sold, also for 40,000 dal. I ventured upon a stroke and bought it, but became indebted for 20,000 daler copper coin; we shall see if or how soon I can disburse this, who must live of a profession, without any emoluments. Now avails neither correspondence, nor anything in *re literaria*. I have laid out from shore, the anchor is weighed. I must sail on, we shall see how soon I reach the haven. 200 tuns of grain I receive in rental, but to maintain four dragoons in time of war also means something."

Regarding a medallion taken of the features of Linnæus, he himself wrote to his friend P. Wargentin.

"A new and great proof of your complaisance towards me I have had these days through Herr Julander, who has moulded my features in wax to such perfection, that all say they never have seen anything so excellent and so like. We see even now with astonishment that men of our people, when they in full earnest devote themselves to anything, go farther than other nations. It would be an irreparable harm, if such an excellent subject should not be enabled to travel, and develop his art to the highest possible degree, for such men form epochs in our era."

In some of the many foreign biographies of Linnæus has been reiterated an unpleasant account of his wife, which one of his pupils had spread about, that the genial Linnæus should have been mastered by his wife, and to such a degree, that she had forced him to commit a great injustice to their son, and who, it was said, was hated by his unnatural mother. It is to be regretted that this calumny has been perpetuated in print, but it was fortunate that the family whom it concerned, knew of nothing but the happiest home life of concord and harmony. Carl von Linné, *Junior* suffered already in his youth from hypochondria. His mother, who managed the entire economy of the house, thought that, considering he at very young years had got a salary from the government, and had secured the professorial chair of Botany for the future, and his father had willed his great and valuable library, and all his great collections, with the exception of the botanical one, to his son as his successor, which Linnæus himself in his testament speaks of as being worth twice as much as all else together, his mother thought that in consideration of all this, young Carl ought not also to share in the division of real property, together with his mother and sisters. It is this which gave birth to the obloquy. But that Linnæus stipulated that his grand Herbarium should be sold to the highest bidder for the benefit of his daughters, is much more open to a question, the justice of which might still be mooted, for in Linnæus's supplication to the government for his son to inherit the professorial chair of Botany, the ever just and clear sighted Linnæus distinctly says that his son was to inherit his library and *all* his collections, including the famous grand Herbarium, valued 100,000 dal. and in consideration of all this, Carl von Linné, junior, did obtain the appointment.

That this great Herbarium was afterwards willed away to the daughters was incontestibly a grievous fault, and can only be accounted for by the feeble condition, approximate to senility, in which Linnæus is said to have been at the time of his making his will, and during the last years of his life.

His wife may, or she may not, have clearly understood that to secure his son's appointment his father had distinctly promised that the future professor of of Botany should inherit the Herbarium as well.

This ultimately led to his valuable collection becoming through puchase the possession of Dr. J. E. Smith, who presented it to the Linnæan Society in London, 1788, of which he was the first President. Linnæus never dreamt that this Herbarium would be lost to Sweden, nor would it ever have been, had not the petty jealousy of Thunberg, his old disciple, prevented the purchase by the Swedish government or the University of Upsala.

Unfortunately Gustavius III. was at this particular period at the baths of Pisa in Italy, or this sale, which afterwards justly has been reckoned a national calamity, had never been allowed to take place. The king, on hearing of the fate of the grand Herbarium, immediately dispatched a man-of-war to overtake the vessel in which it was brought to England, but the expedition mis-carried, and thus London to-day boasts of possessing Linnæus's grand botanical collection.

The Swedish government has tried to console itself by buying the estate of Hammarby, and maintaining it at its expense, as a national property, thereby honouring the name and memory of the great son of Sweden, but it is like treasuring the empty casket when the jewel is gone, which might not have passed into strange hands. Sir Hans Sloane had also stipulated in his will that his collection should be offered to many British and foreign Universities and Societies for sale, but England, ever to the fore when the question concerns a great national honour and benefit, secured Sir Hans Sloane's collections, and built for them the British Museum!

The cause of the different fate which befel the two grand collections of the two respective countries, is to be found in that national trait of petty jealousy which has thwarted so many brilliant aims by eminent Swedes, of which the history of that country affords abundant proofs, past and present.

That Linnæus dearly loved all the members of his family is not denied, even by those who would insinuate that he had been led to commit an injustice to his son, which unconsciously instead turned upon his country.

That he took great pride in his son may be inferred from his annotations, in which, after speaking about the many acts of favour shewn him by the brilliant reigning Queen Lovisa Ulrica, he says, "But the greatest joy for Linnæus was that Her Majesty, the incomparable Queen, asked after Linnæus's only son, and what capacity and zeal he had for Natural History, and when she learnt that, she promised, that when he had grown up, she would allow him a free journey, at her expense, all over Europe, which gracious promise gladdened Linnæus at heart."

A younger son, Johan, died in infancy, and a grown-up daughter died also before him, but three of his daughters survived their famous father. The eldest of these children had inherited much of her father's love of nature. She made the discovery in the garden at Hammarby, that a faint electric light sometimes flashes from certain plants late of an evening, a curious fact which might with great appositeness be studied in this our age of electricity. Poets have in all ages endowed flowers with souls, what if electricity—which by some scientist is called the soul of the universe—also permeates every little herb, as most things else? What more likely? This slight hint, left us as a legacy from the last century, might be worthy the investigative spirit of our own time.

The youngest daughter seems to have been the apple of her father's eye, perhaps, because when she was born there was scarcely any sign of life in her, but her anxious father hastened to blow his breath into her lungs, and by that means he succeeded in calling her to life. These two daughters married, but the second daughter, Lovisa, continued to reside with her widowed mother at Hammarby, and both ladies reached a great age, the daughter did not die till 1839, at ninety years of age, the last survivor of the famous name of Linnæus or Linné.

Carl von Linné junior, who on succeeding to the professorship seemed to be expected to fill the place of him, whose loss was irreparable, felt more dejected than glad of his responsible position.

He truly loved the Science to which he, like his great father, devoted himself, but the electric spark of genius had not kindled in his soul. People generally expect too much from a great man's son, and he felt depressed that such should be the inconsiderate demand by the world, and his melancholy increased. He was fated not to hold the inherited position long, for shortly after a journey to France and England, on his return home, he died in 1783, and with him died the male line of the ennobled family name of Linné.

CHAPTER XVIII.

ANY scientific societies have been instituted in honour of Linnæus, the foremost one being that of London, founded March 18th, 1788, by Sir Joseph Banks, Bishop Sam. Goodenough, J. E. Smith, M.D., and others, and of which J. E. Smith was made president for life. The president of 1888 being W. Carruthers, F.R.S., F.G.S.; the preceding was Sir John Lubbock. This Society, which in 1856, removed from Soho Square to Burlington House, Piccadilly, publishes yearly "The Botanical Journal," "The Zoological Journal," "'Transactions' and 'Proceedings' of the Linnæan Society."

Societé Linnéenne was instituted the same year in Paris as in London; in Philadelphia, 1806; in Boston, 1813; in Bordeaux, 1818; also in Lyons, and in New South Wales, Australia, exist similar Societies.

Linnæus's writings were for many years prohibited in the pontifical states, but in 1773, Cardinal de Zelanda, caused professor J. F. Maratti, to be succeeded by Ant. Minasi, purposely to lecture about Linnæus's "Systema Sexuale."

The portrait of Linnæus in the dress of a Laplander, was painted in 1737, in Holland, by Mart. Hoffman, of which an engraving forms the frontispiece of this volume. On Linnæus's tomb of black porphyry in the cathedral of Upsala, is a medallion portrait by the eminent Swedish sculptor Sergel, contemporaneous with Linnæus, and is deemed an admirable likeness. The remains of Linnæus were interred, January 22nd, 1778.

A bust of Linnæus in biscuit, by John Forslund, was in 1807 placed in the academy of Wexiö, the school which Linnæus frequented as a boy.

A marble statue, in sitting posture of Linnæus,

modelled by Byström, was erected October 22nd, 1822, in the Botanical Garden at Upsala, where Linnæus used to lecture.

A small obelisk was on June 12th, 1866, erected in front of the cottage at Råshult, to commemorate Linnæus's birth at that place, May 13th, 1707. This obelisk replaced one of a more primitive nature, which was deemed unseemly, and had been erected by some local admirer. The present one is seen as the passenger hurries past on the railway, skirting the child-Linnæus's garden plot.

A magnificent bronze statue of Linnæus, modelled by professor Frithiof Kjellberg, was unveiled May 13th, 1885, at Stockholm, the event being held a national fête. It stands at the back of the Royal National Library, and central in the park which has been renamed the Linnæan Park. The statue itself is surrounded by four allegorical female figures, those of Botany, Zoology, Mineralogy, and Medicine. An engraving thereof adorns this volume.

London has also honoured Linnæus, by placing a life-sized statue in a niche at the back of Burlington House, the left frontage of which contain "The

Monument to Linnaeus at Stockholm.

Linnæan Society," and many interesting portraits and relics of the famous Swedish Naturalist.

The late Dr. Ewald Ährling summed up as follows:—" It is generally recognised as the principal merit of Linnæus, that he established the sexual system, or the scientific arrangement of the various form of the vegetable kingdom according to the sexual parts of the flowers : stamens and pistils. Although the method, from a practical view, is of particular significance, as it can be easily learned, and utilized, yet purely scientifically it becomes of less moment than many others. To reach—so to speak—the very life of the plants, we must quite naturally follow their development from seed to seed.

"This of course is not possible, if we exclusively take into consideration some certain parts of the plant, during a certain period of the development. Linnæus perceived and acknowledged this sooner than anyone else; but he also shows that the ideal of the systematizing part of the science is to seek to approach this goal, but that the same never can be reached by any mortal. For the great majority of his pupils, the sexual system was sufficient. He initiated only a very few into his '*ordines naturales*,' whose value has only been rightly understood and appreciated by a much later time. Linnæus evinced quite as great merit in that he particularly incited to biological studies, purified the language, fixed the terminology, founded the character of family, introduced art names instead of the former long descriptions. The necessity of the latter he compared himself with putting a tongue in a bell. With Linnæus was begun quite a new era in the history of Natural Sciences, for so thoroughly was, and became, his reform."

But the eminent scientist Dr. J. Sachs in his work "Geschichte der Botanik von 16 Jahrhundert bis 1860,"—(München 1875,) has attacked Linnæus as having only followed other experts in classifying, and in the art of characterising, relegating Linnæus to being solely a systematizer and not a physiologist as well. However, Dr. J. G. Agardh, in a paper which he read at the Centenary Celebration of Linnæus's death, January 10th, 1878, in Lund, ably refutes the would-be detractor of

Linnæus's work, for every man must be judged according to the times in which he lived and worked, and particularly so a scientific man, when every century produces new progressive systems that supersede one another, the products of the evolution of time.

Dr. Sachs maintains that Rued Jac Camerarius's sexual system had long preceded Linnæus's, and so it had, but many observations had been made, then as hastily forgotten, and long afterwards been brought forth as new discoveries in science, and Agardh reminds us that "already the ancients knew the circumstances that necessitated male flowers to exist in the proximity of the female so as to obtain fruits from the date-palm, the cultivated fig-tree, etc." and that "already four years before Linnæus became a student at Lund he had in the garden of Stenbrohult made experiments with a cucumber-plant, which he had deprived of its male flowers, and then found that the female flowers did not fructify." Linnæus himself refers to the writings of Camerarius, and to the *sexu plantarum* of Vaillant which he read in 1729, as well as "The Nuptials of

Trees," by G. Wallin, December 1729, and which the late Dr. Ährling considers the immediate impulse to "*Preluda sponsalioum plantarum*," and "*Methodus propria et nova a sexu desumta.*"

The reader has himself on pages 31 and 32 of this little book read what Linnæus says about the acquatic plants, and that already *Michelins* had noticed the same, but that yet the sexual system had not struck him. Of course in science as in everything else, many different people at vastly different times may conceive similar ideas. But hear what Linnæus himself says, anticipatory, by way of a reply to the learned German doctor:—"it is difficult to say who is the actual discoverer of the sexuality of plants, for it is the case with the majority of discoveries as it is with rivers, which begin with small tributaries from various sources, until at last they gain such strength that they are able to carry onward even the heaviest loads." And again, we may refer to a page of this book, 167, where Linnæus writes "that's the reward here in Sweden, where after the manner of the Germans, we labour to refute that which we do not ourselves understand,

and hate what is in any way remarkable, because some one else has first observed it."

But all things changeth, reputations are built up, and by a succeeding generation ruthlessly pulled down, the Linnæan Era has passed, the Herbarium which Linnæus thought so vast, is twenty-folded by the knowledge of to-day; his system is superseded, and his collections have dwindled away until they are held in merely a few drawers, that principally contain shells and insects, the latter much damaged by time, that being all which is yet preserved in the Linnæan Society of the Natural History collection which Dr. J. E. Smith bought of Linnæus's heirs for 900 guineas, and September 29th, 1784, brought away in twenty-six boxes in the vessel "Appearance." But already in March 1796, the collection of mineralogy was dispersed by auction. And all the natural objects preserved in spirits of wine exist no more; the mammals have disappeared, the birds have flown, and the fishes slipped away, probably caused by the several removals, first to Dr. J. E. Smith's private museum, and afterwards to and from the location in Soho Square. Together with Dr. J. E.

Smith's own collection they had been sold to the Linnæan Society for £5,000. But in its archives part of Linnæus's vast correspondence still exists, more than a thousand letters, mostly in Latin and Swedish, and although several collections of letters of Linnæus have, from time to time, been published in English, German, French, Italian, and Swedish, comprising the correspondence which was carried on with the *savants* of these respective countries, nearly numbering 5000 letters; still, as only 708 out of 1253 known to be from Linnæus himself have been published, a great number must still be waiting for an able hand to arrange them, and give a choice selection of this hidden treasure to the world at large, for as a recent Swedish critic has well said:—"Linnæus, also on account of his oratorical and literary merits well deserves his laurels, for his diction falls like a toga around the divine form of his thoughts. The naïvely simple tone which pervades it, accords well with that pious and child-like heart, which beats, almost audibly, between the lines."

FINIS

www.ingramcontent.com/pod-product-compliance
Lightning Source LLC
Chambersburg PA
CBHW020759230426
43666CB00007B/760